PERGAMON SCIENCE SERIES

ELECTRONICS AND WAVES—*a series of monographs*
EDITOR: D. W. Fry (Harwell)

MILLIMICROSECOND PULSE
TECHNIQUES

MILLIMICROSECOND PULSE TECHNIQUES

By

I. A. D. LEWIS

M.A. (Cantab.), A. Inst. P., Graduate I.E.E.

(formerly Assistant Lecturer in Physics at the University of Liverpool; now at the Atomic Energy Research Establishment, Harwell, Berkshire)

AND

F. H. WELLS

M.Sc. (Eng.), D.I.C., A.M.I.E.E., S.M.I.R.E.

(Atomic Energy Research Establishment, Harwell, Berkshire)

NEW YORK: McGRAW-HILL BOOK CO., INC.
LONDON: PERGAMON PRESS LTD.
1954

*Published in Great Britain by Pergamon Press Ltd., Maxwell House, Marylebone Road,
London, N.W.1.*
Published in U.S.A. by McGraw-Hill Book Co., Inc., 330 West 42nd Street, New York 36, N.Y.

Made in Great Britain at the Pitman Press

CONTENTS

	PAGE
EDITOR'S PREFACE	xi
AUTHORS' PREFACE	xi
ACKNOWLEDGMENTS	xiv
1 THEORETICAL INTRODUCTION	1
1.1 The Laws of Circuit Analysis	1
1.2 Sinusoidally Varying Currents	2
1.3 Fourier Analysis of Pulse Waveforms	4
1.4 The Unit Step Function	7
1.5 Laplace Transforms	9
1.5.1 Basis of Method	9
1.5.2 Approximations	11
1.6 Simple Variable Circuit	14
1.7 Circuits with Distributed Parameters	15
2 TRANSMISSION LINES	17
2.1 Introduction	17
2.2 Uniform Rectilinear Lines	17
2.2.1 Summary of Properties	17
2.2.2 Analysis	19
2.2.3 General	22
2.2.4 Terminations and Discontinuities	22
2.2.5 Transmission line as a Circuit Element	32
2.2.6 Losses	38
2.3 Helical Lines	42
2.3.1 Formula for Inductance per Unit Length	42
2.3.2 Capacitance per Unit Length	47
2.3.3 Phase Distortion	48
2.3.4 General	51

	PAGE
2.4 Lumped Delay Lines	52
2.4.1 Constant-k Filters	53
2.4.2 Derived Filters	55

3 TRANSFORMERS 59

3.1 Introduction	59
3.2 Elementary Matching Networks	60
3.3 Lumped Pulse Transformers	61
3.3.1 Equivalent Circuit	61
3.3.2 Limitations on Performance	62
3.4 Tapered Lines	63
3.4.1 Quarter-wave Transformer	63
3.4.2 Analysis of Smooth Tapered Transmission Lines	64
3.4.3 Gaussian Line	76
3.4.4 Exponential Line	79
3.4.5 Linearly Tapered Coaxial Line	90
3.4.6 Other Laws of Impedance Variation	91
3.5 Transmission-Line Type of Pulse Inverter	91
3.5.1 Principle of Operation	92
3.5.2 Helical Line Arrangement	94
3.6 Coupled Line Transformers	96

4 PULSE GENERATORS 99

4.1 Introduction	99
4.2 Discharge Line Type Pulse Generators	100
4.2.1 Mechanical Relays	102
4.2.2 Thyratron Pulse Generators	107
4.2.3 Tapered Discharge Line	112
4.3 Pulse Generators Employing Secondary Emission Valves	115
4.3.1 Simple Trigger Circuit	116
4.3.2 Practical Circuit	118

			PAGE
4.4	Further Possibilities		120
4.5	Attenuators		121
	4.5.1	Limitations of Resistors at High Frequencies	121
	4.5.2	Simple Lumped Attenuators	122
	4.5.3	High Frequency Improvements	124
	4.5.4	Lossy Transmission Line Types	126
4.6	Reflex Peak Valve-voltmeter		127
	4.6.1	Principle of Operation	128
	4.6.2	Blocking Oscillator Transformer Construction	130

5	AMPLIFIERS		132
5.1	Introduction		132
5.2	Properties of Valves		132
	5.2.1	High Frequency Limitations	133
	5.2.2	Valve Requirements—Figure of Merit	135
	5.2.3	Some Valve Types	138
5.3	Interstage Coupling in Cascade Amplifiers		138
	5.3.1	Practical Circuit	139
	5.3.2	Use of Dynode in Secondary Emission Valves	141
	5.3.3	General	142
5.4	Distributed Amplifiers		143
	5.4.1	First Order Theory	145
	5.4.2	Further Considerations	151
	5.4.3	Practical Circuits	161
5.5	Special Valves		167
	5.5.1	Transmission-Line Tubes	167
	5.5.2	Travelling-Wave Tubes	168

6	CATHODE RAY OSCILLOSCOPES		170
6.1	Introduction		170
6.2	Cathode Ray Tube Design		170
	6.2.1	Transit-Time Limitations	171

		PAGE
6.2.2	Connections to Deflecting Plates	174
6.2.3	Overall Frequency Limitations of Normal Deflecting Plate Structures	177
6.2.4	Methods of Reducing Display Distortions due to Deflecting Plates	181
6.2.5	Spot Size and Deflection Sensitivity	187
6.2.6	Brightness	191
6.2.7	Beam Acceleration After Deflection	193
6.2.8	Sealed versus Unsealed Cathode Ray Tubes for Photographic Recording	194
6.2.9	Photographic Technique	195
6.2.10	Performance of Some Cathode Ray Tubes	196
6.3	Circuit Design for Transient Recording Oscilloscopes	196
6.3.1	Signal Delay Circuit	197
6.3.2	Time-Base Circuits	198
6.3.3	Auxiliary Circuits	205
6.3.4	Camera Equipment	207
6.4	Oscilloscopes for the Display of Recurrent Waveforms	207
6.5	Oscilloscopes for Recurrent Pulses Using Pulse Sampling Techniques	208
6.5.1	Mixing Circuits	210
6.5.2	Frequency Limitations depending upon Sampling Pulse Duration	212
6.5.3	Display Circuits	215
6.5.4	Brightness of the Display	217
6.5.5	General	217
7 APPLICATIONS TO NUCLEAR PHYSICS		218
7.1	Introduction	218
7.2	General Measurement Problem	218
7.3	Scintillation Counters	219
7.3.1	Current Pulse Shape	220

	PAGE
7.3.2 Performance of Existing Scintillation Counters	223
7.3.3 Spread of Transit-Time in Photomultiplier Tubes	224
7.3.4 Scintillation Counter Output Circuit	225
7.3.5 Particle Counting by Čerenkov Radiation	229
7.3.6 Pulse Testing of Photomultipliers	229
7.4 Spark Counters	231
7.5 Amplitude Discriminators	232
7.5.1 Use of Amplitude Discriminators	238
7.6 Fast Scaling Circuits	239
7.7 Coincidence Circuits	244
7.7.1 Pulse Limiter with Diode Mixer Circuit	246
7.7.2 Pulse Amplitude Selection for Coincidence Units	253
7.7.3 Stability of Coincidence Circuits	255
7.7.4 Factors Determining the Minimum Possible Resolving Time	256
7.7.5 Mixer Circuits	258
7.8 Time Interval Measurements with Delayed Coincidence Circuits	264
7.8.1 Time Sorters	269
7.8.2 The Chronotron Timing Unit	269
7.9 Time Interval Measurement by Integration Methods	271
7.10 Recording Oscilloscope Measurements	274
8 MISCELLANEOUS APPLICATIONS	**276**
8.1 Introduction	276
8.2 Millimicrosecond Pulse Generators for Narrow Bandwidth Radio Receiver Measurements	276
8.3 The Use of Millimicrosecond Pulses for Radar Propogation Measurements	277
8.4 The Investigation of Electrical Discharge Phenomena with Transient Recording Oscilloscopes	277

	PAGE
8.5 Electro-Optical Shutters for High-Speed Photography	278
8.5.1 Kerr Cell	278
8.5.2 Image Converter	280
8.6 Conclusion	282
APPENDIX I. The General Distortionless Transmission Line	283
APPENDIX II. Transmission Line Characteristic Impedances	286
APPENDIX III. Some Valve Data	290
APPENDIX IV. Guide to Notation	291
BIBLIOGRAPHY	293
REFERENCES	294
INDEX	308

EDITOR'S PREFACE

THE aim of these monographs is to report upon research carried out in electronics and applied physics. Work in these fields continues to expand rapidly, and it is recognized that the collation and dissemination of information in a usable form is of the greatest importance to all those actively engaged in them. The monographs will be written by specialists in their own subjects, and the time required for publication will be kept to a minimum in order that these accounts of new work may be made quickly and widely available.

Wherever it is practical the monographs will be kept short in length to enable all those interested in electronics to find the essentials necessary for their work in a condensed and concentrated form.

D. W. FRY

AUTHORS' PREFACE

THE techniques of applied electronics are concerned with the handling of information by electrical devices. Systems in use fall fairly definitely into two classes employing either modulated continuous wave techniques, or, alternatively, pure pulse methods. The first group includes the familiar transmission of intelligence by radio or carrier telephony and in the second class are to be found radar timing circuits, for example, electronic digital computers and counting equipment for use in nuclear physics research. The concept of frequency bandwidth, originally introduced in connection with the spectrum of frequencies associated with a modulated C.W. signal, is also applied to pure pulse systems. The speed and accuracy with which information may be handled is determined, in both classes, by the bandwidth of the system and the historical development of the subject, from line telegraphy to modern pulse modulated microwave links, indicates a continual striving after larger bandwidths and shorter and sharper pulses.

We may define the millimicrosecond range as that in which time intervals lying between 10^{-6} and 10^{-10} sec. are of interest. The former figure is chosen as representative of the state of development of pulse techniques as reached at the end of World War II and the latter is an optimistic lower limit set by physical considerations. The corresponding frequency band involved extends from 1 Mc/s, or less, up to 10,000 Mc/s, a limit of 1000 Mc/s being a more realistic goal at the present stage of development. This range covers the field of all short wave radio right up to the microwave region but we are only concerned, as the title of the book indicates, with circuit techniques employed in systems of the second class; short wave radio methods are not directly of interest except in so far as they provide background knowledge and experience in working with high component frequencies. Modulated C.W. microwave systems, having bandwidths of say 1000 Mc/s centred on a frequency of 10,000 Mc/s, may be developed in the future and the "video" channels of such systems would fall within our range of interest. Important developments are taking place in this direction but we shall not be concerned here with any such complete equipments.

In the microsecond region the familiar circuit parameters of resistance, inductance and capacitance may be treated as discrete entities implying that the linear dimensions of the apparatus are very much shorter than the wavelength of electromagnetic radiation corresponding to the highest frequency components involved; electron transit time effects in thermionic vacuum tubes are negligibly small. In the microwave region, at the other extreme, linear dimensions are of prime importance, as for example in waveguide circuits and cavity resonators, and electron transit time phenomena form the basis of operation of the tubes employed. The millimicrosecond range lies in the interesting region between the two in which the techniques involved tend to overlap. Thus on the one hand, we try to improve conventional pulse circuits and push these to the limit of speed and, on the other, develop arrangements in which the circuit parameters are deliberately distributed in space and which possess wave propagation properties. Our theme is the development of devices of large bandwidth, extending down to comparatively low frequencies (1 Mc/s) and we are concerned, for example, with transmission lines operating in the principal mode but not with waveguide structures which are dispersive. It is found that ordinary circuit theory applies to such distributed circuits but we must be prepared in cases of doubt to fall back on the more general theory of electromagnetic waves.

A glance at the table of contents shows the scope of the book. A brief theoretical introduction is included for the benefit of the non-electronic physicist and to clarify terminology. The bulk of the work is devoted to a consideration of basic circuit elements and pieces of equipment of universal application. Details of specific applications, mostly in the field of nuclear physics instrumentation, fill the last two chapters. A short Bibliography and a comprehensive list of references complete the volume.

It is hoped that the book may be of use to the physicist who, with perhaps little experience in the electronic art, wishes to call these new techniques to his aid. For the electronic engineer, the volume aims at the collation of relevant material taken from known fields of electronic engineering, together with an account of special developments in the millimicrosecond range.

The authors wish to make acknowledgment to the Director, Atomic Energy Research Establishment, for permission to publish the results of some of their work.

I. A. D. L.
F. H. W.

ACKNOWLEDGMENTS

THE authors and the publishers wish to express their thanks to all those who kindly gave their permission for the reproduction of copyright material in this book. Details of sources are as follows:

Proceedings of the Institution of Radio Engineers: Figs. 2.8 and 2.9 from WHINNERY et al. 1944, p. 697; Fig. 2.10 (e) from PETERSON 1949, p. 1294; Fig. 3.9 from SCHÄTZ and WILLIAMS 1950 **38** 1209 and Fig. 6.12 from LIEBMANN 1945 **33** 381.

Electronic Engineering: Figs. 4.9 and 4.10 from MOODY 1952, p. 218; Figs. 5.2, 5.3, 5.4 and 5.7 from May 1952 issue; Fig. 6.23 from McQUEEN 1952 **24** 436; Figs. 7.11 and 7.12 from MOODY 1952 **24** 214; Figs. 7.26 and 7.27 from McLUSKY and MOODY 1952 **24** 330; Figs. 7.40 and 7.43 from MOODY 1952 **24** 289.

Review of Scientific Instruments: Figs. 5.12, 5.13 and 5.14 from KELLEY 1950 **21** 71; Fig. 7.5 from SMITH 1951 **22** 166; Fig. 7.16 from FITCH 1949 **20** 242; Fig. 7.29 from BAY 1951 **22** 397; Fig. 7.31 from DICKE 1947 **18** 907; and Fig. 7.33 from HOFSTADTER and McINTYRE 1950 **21** 52.

Journal of the British Institution of Radio Engineers: Fig. 4.7 from WELLS 1951 **11** 495; Fig. 6.13 from DUDDING 1951 **11** 455; Figs. 8.2 and 8.3 from JENKINS and CHIPPENDALE 1951 **11** 505.

Journal of Scientific Instruments: Fig. 6.18 from PRIME and RAVENHILL 1950 **27** 192; Figs. 7.14, 7.15, 7.17 and 7.18 from WELLS 1952 **29** 111.

Canadian Journal of Research: Figs. 7.28 and 7.36 from BELL, GRAHAM and PETCH 1952 **30** 35.

Trans. Amer. Institution of Electrical Engineers: Fig. 4.13 from DAWES et al. 1950 **69** 571.

Nucleonics: Figs. 4.11, 6.20, 7.7 and 7.10 from WELLS 1952 **10** 28.

The curves in Fig. 1.3 were copied by permission from *Radio Engineer's Handbook*, by TERMAN. Copyright, 1943, McGraw-Hill Publishing Company Ltd.

Electronics: Fig. 7.4 from RAJCHMAN and SNYDER 1940 **13** 20; Appx. II diagrams from BARCLAY, issue of Feb. 1942, p. 50.

Figs. 4.16 and 6.17 are reproduced by permission from M.I.T. Laboratory for Insulation Research, Tech. Report No. 20 (FLETCHER) and a separate paper by D. O. SMITH respectively. Fig. 6.16 is from an article by BAYER and NETHERCOTT in British Electrical & Allied Industries Research Association Report U/T115.

1

THEORETICAL INTRODUCTION

This chapter is intended to serve as an introduction for the non-electronic physicist who may wish to be reminded of some of the terminology employed by circuit engineers and to whom a review of circuit theory, though superficial, may be of assistance (see ELMORE [101]). Definitions given are not expressed particularly rigorously but are included to clear up ambiguities and to avoid repetition of phrases throughout the main text.

1.1 THE LAWS OF CIRCUIT ANALYSIS

KIRCHHOFF's Laws provide the foundation upon which all circuit theory is based. In the first place we are concerned with lumped circuits, that is, circuits composed of discrete entities (resistors for example) connected together by wires. OHM's Law introduces the idea of a linear circuit element; the potential difference across it is directly proportional to the current flowing through it.

When applied to the several junctions and loops in any network these laws give rise to a number of simultaneous equations (called the network equations) which are linear in the voltage and current variables provided the circuit elements are linear. In our theoretical analyses we shall assume, unless otherwise stated, that all circuit elements are linear. The Principle of Superposition follows at once; thus if an e.m.f. E_1 of any character whatsoever, acting at one point (termed the input) produces a current I_1 at any point (termed the output) and a second e.m.f. E_2 at the input causes a current I_2 to flow in the output then an input e.m.f. equal to $E_1 + E_2$ will produce an output current $I_1 + I_2$.

The labour of solving the network equations can often be reduced, or avoided altogether, by successive application of the elementary rules indicated in fig. 1.1 and by making use of THEVENIN's Theorem. This theorem states that any two terminal network of linear resistors and sources of e.m.f. is equivalent to a single e.m.f. E in series with a single resistor R. Any two junction points A and B may be chosen as the relevant two "terminals" and R is termed the internal resistance or output resistance of the network (with respect to AB). The e.m.f. E is equal to the potential difference across AB, before any external circuit is connected thereto, and the equivalent resistance is equal to that seen on looking into the two terminals with the sources of e.m.f. inoperative but left connected. R is also

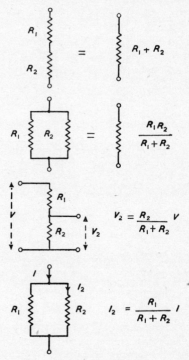

Fig. 1.1. The elementary rules of circuit analysis.

given by the ratio E/I where I is the current which would flow in a short circuit connected externally across AB. The two forms of equivalent circuit are shown in fig. 1.2; the source of e.m.f. E, or constant voltage generator, has zero internal resistance, and the constant current generator I has infinite internal resistance.

1.2 SINUSOIDALLY VARYING CURRENTS

The principles outlined above apply, in the first instance, to circuits involving direct currents only. When time varying currents are considered KIRCHHOFF's laws must still be satisfied and, in addition to OHM's law for a resistance, we have the fundamental relations

$$V = L \cdot \frac{dI}{dt} \qquad (1.1)$$

for an inductance L,* and

$$V = \frac{1}{C} \int I \cdot dt \qquad (1.2)$$

for a capacitance C.

These equations are linear (if ideal capacitors and inductors are envisaged) provided the values of L and C (and also of R for that

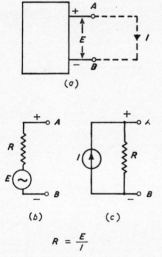

$$R = \frac{E}{I}$$

Fig. 1.2. (b) Constant voltage source equivalent to (a).
(c) Constant current source equivalent to (a).

matter) do not vary with time. We shall restrict ourselves completely, unless otherwise stated, to circuits in which the parameters are invariable.

The network equations are now a set of simultaneous linear ordinary differential equations, with constant coefficients. They are

---

* This relation is first met with in the form $E = -L \cdot \dfrac{dI}{dt}$ but it is preferable to regard an inductor as an impedance of some sort across which is developed a potential difference due to the current through it rather than as a source of e.m.f. in the circuit.

satisfied, for example, if all variables change sinusoidally with time; the equations then reduce to algebraic equations involving the various amplitudes and phase angles. The application of complex algebra enables us to arrive directly at the latter equations without first writing down the differential equations. A voltage V, for example, which varies sinusoidally at an angular frequency ω having an amplitude V_0 and a phase angle θ may be written as the real part of the complex quantity

$$V_0 e^{j(\omega t + \theta)} = \bar{V} e^{j\omega t}$$

where $\bar{V} \equiv V_0 e^{j\theta}$ is the complex amplitude which includes both the actual amplitude and the phase. On substituting complex variables in equations 1.1 and 1.2* the time factor disappears and the voltage and current complex amplitudes are related in a manner analogous to Ohm's law. The reactances $j\omega L$ and $-j/\omega C$ play the same rôle as resistance R. Kirchhoff's laws continue to apply to the complex amplitudes and the familiar rules of fig. 1.1 may be used provided the complex reactances $j\omega L$ and $-j/\omega C$ (and R) and the complex amplitudes of the e.m.fs., potential differences and currents are employed. Thevenin's theorem again holds, but if e.m.fs. of several different frequencies are present the effect of the network on an externally connected circuit (or vice-versa) must be worked out using a separate equivalent, for each frequency. The resultant current or voltage is then obtained by addition.

1.3 FOURIER ANALYSIS OF PULSE WAVEFORMS

We are interested in calculating the behaviour of a circuit when the input is in the form of a pulse rather than an infinite train of sinusoidal oscillations. Fourier's Theorem states that any function of time is equivalent to a sum of continuous sine waves of frequencies extending from zero to infinity and of suitable amplitudes and phases. If the function is periodic the frequency spectrum consists of a number of discrete lines, harmonically related, and the amplitude of each component is finite. For an isolated pulse the spectrum becomes continuous; the amplitude of each component tends to zero, being replaced by an amplitude per unit frequency range and

---

* The constant of integration represents a superimposed steady voltage (d.c. level) which does not affect the a.c. behaviour and may therefore be put equal to zero.

the sum is replaced by an integral. If $f(t)$ is the (real) function of time representing the pulse then the relative complex amplitude $A(\omega) \cdot d\omega$ of angular frequencies lying between ω and $(\omega + d\omega)$ is given by

$$A(\omega) = \int_{-\infty}^{\infty} e^{-j\omega t} f(t) dt \qquad (1.3)$$

The frequency spectra of a single rectangular pulse and of a symmetrical triangular pulse, of the same duration and double the peak amplitude, are compared in fig. 1.3. The greater importance of

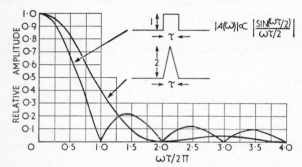

Fig. 1.3. Fourier spectra of rectangular and triangular pulses.
(Curves copied by permission from *Radio Engineer's Handbook*, by TERMAN. Copyright, 1943, McGraw-Hill Publishing Company, Ltd.)

the high frequency components in the case of the rectangular pulse, due to the shorter rise-time of the latter, is evident.

The first step is to analyse the input pulse into its Fourier components and calculate separately the corresponding components of the output. By the Principle of Superposition the resultant output is then given by the integral (or sum) of the individual components. Thus, if a sinusoidal input of amplitude $A(\omega)$ gives an output $G(\omega) \cdot A(\omega)$, where $G(\omega)$ is characteristic of the network, then the output pulse $\phi(t)$ corresponding to the input $f(t)$ is

$$\phi(t) = \frac{1}{2\pi} \int_{-\infty}^{\infty} G(\omega) A(\omega) e^{j\omega t} d\omega \qquad (1.4)$$

This procedure involves two integrations, often of the contour type, whereas the method of Laplace Transforms, discussed in the next section, usually requires only one such integration. The latter method is more convenient in pulse work but the concept of frequency components, inherent in the Fourier approach, is exceedingly useful and enables approximate estimates of performance to be rapidly made.

In order that an input pulse may be reproduced without distortion (a) the circuit must amplify (or attenuate) all frequency components to the same extent and (b) the phase change suffered by each component on passing through the circuit must either be zero or directly proportional to the frequency (a simple time delay without distortion of the pulse shape being permissible). Every practical system suffers from *amplitude distortion*,* in which case (a) is not satisfied, and also from *phase distortion* in which event the condition (b) is not met. Any system possesses a certain *amplitude bandwidth* of frequencies over which amplification is sensibly independent of frequency, falling to zero outside the band, and also a certain *phase bandwidth* over which the phase change is directly proportional to frequency. It is easily seen that the rise-time of the leading edge of an output pulse, corresponding to a rectangular shaped input pulse, cannot be much less than one quarter of the period of the upper frequency limit of amplitude response of the system. Further, the total duration of the output pulse cannot be much longer than the half-period of the low-frequency limit. The importance of the, perhaps less familiar, concept of phase bandwidth must be emphasized in pulse circuits. It is preferable, from the point of view of obtaining minimum distortion, that frequency components should not be reproduced at all rather than arrive at the output in the wrong phase. In the latter case, the high frequency components do not add correctly to give a rapid pulse rise but merely produce an undesirable series of "wriggles" which appear superimposed on the leading edge of the output pulse: these may occur shortly before and up to the time at which the main pulse arrives and may also persist along its top. The effect is easily recognizable but difficult to eliminate; since the phase and amplitude bandwidths are intimately related it is not always easy to arrange for the amplitude response curve to "cut off" before the phase characteristic departs from linearity.

Much theoretical work has been done on the relationships between bandwidths and pulse distortion, including discussions on the best method of defining the rise-time, and the reader may consult

---

* The term "frequency distortion" is often employed to indicate what is meant here by "amplitude distortion", the latter term being associated with departures from linearity at any given frequency.

papers by CHENG [102], EAGLESFIELD [103], SAMULON [104], DI TORO [105] and TUCKER [106] for further information. A simple practical measure of the time of rise is to take the time interval between the points corresponding to 10% and 90% of the final amplitude. The rise-time t_r is then determined approximately by the relation

$$t_r = 0.4/f_c \qquad (1.5)$$

where f_c is the upper frequency limit to the uniform response band defined by the point at which the amplitude falls to a fraction $1/\sqrt{2}$ of (or is 3 db down on) its medium frequency value.

When a number of networks are connected in cascade the amplitude-frequency response curves are multiplicative (or additive when a decibel scale is employed) and the phase-frequency curves are additive. We are concerned with the cumulative deterioration in pulse rise-time which occurs on passing through the whole system and the following approximate relation is invaluable. If the quantities t_r are the rise-times of the individual sections, when a rectangular pulse is applied to each separately, then the overall rise-time τ of the whole is

$$\tau = \sqrt{\Sigma t_r^2} \qquad (1.6)$$

The low frequency limit of the system, which determines the maximum pulse duration which may be handled, usually gives us comparatively little cause for concern, except in the case of transformer design and in certain types of travelling-wave amplifier tubes. The limit lies well within the microsecond range and "conventional" circuit techniques may be adopted to make it as low as we please; the measures taken must not, of course, interfere indirectly with performance at very high frequencies which is our primary concern in the millimicrosecond region.

1.4 THE UNIT STEP FUNCTION

In pulse work the Heaviside step function is a more convenient basic waveform, both for theoretical and experimental investigations, than is a train of sine waves. The unit step function, written $[1]_0$, has the value zero for all times up to $t = 0$ and the value unity thereafter. Any waveform may be synthesized as the sum of a number of step functions of suitable amplitudes (and signs) delayed by suitable amounts. A rectangular pulse, for example, of height A

and duration τ commencing at time $t = 0$ is equivalent to $A \cdot ([1]_0 - [1]_\tau)$ where the suffixes indicate the respective times at which the step functions occur. The Principle of Superposition now shows us that the resultant output is the sum of the individual outputs due to the various step functions of which the input waveform is composed. If $h(t)$ is the output response for a unit step

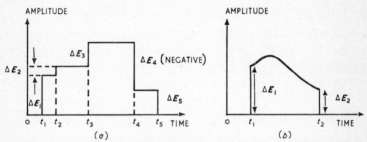

Fig. 1.4. (a) Pulse composed of a limited number of step functions. (b) Pulse with smooth top.

function input at $t = 0$ then the output, at time t, corresponding to an input of the form shown in fig. 1.4(a) is

$$\sum_i h(t - t_i) \Delta E_i \qquad (1.7)$$

where the summation is taken over all the discontinuities. When part (or all) of the input pulse consists of a smooth curve the amplitudes of the steps, and the time delays between each, all tend to zero. The sum becomes the Duhamel Superposition Integral, taken over the smooth curve, and the output at time t is given by

$$\int_0^t h(t - t') dE(t') + h(t - t_1) \Delta E_1 - h(t - t_2) \Delta E_2 \qquad (1.8)$$

for $t_1 < t' < t_2$. For generality, we have here retained two discontinuities as shown, for example, in fig. 1.4(b). The integral may also be expressed in terms of the slope of the input curve as

$$\int_0^t h(t - t') \frac{dE(t')}{dt'} dt' \qquad (1.9)$$

for $t_1 < t' < t_2$.

The step function method has the outstanding advantage over the sine wave approach that the output, for any input, can be

THEORETICAL INTRODUCTION 9

determined when the response to a single step function is known (either theoretically or experimentally). When FOURIER's theorem is to be applied, on the other hand, the response, both in amplitude and phase, has to be determined over a wide range of frequencies.

1.5 LAPLACE TRANSFORMS

We turn now to the method of Laplace Transforms which is particularly suited to an analysis in terms of step functions though it is equally applicable to an arbitrary input waveform (without invoking DUHAMEL's integral).

1.5.1 Basis of Method.

The Laplace Transform $\bar{f}$ of a function of time $f(t)$ is defined by

$$\bar{f} = \int_0^\infty e^{-pt} f(t) dt \tag{1.10}$$

where p is a positive constant. We shall suppose throughout unless otherwise stated, that $f(t) = 0$ for all values of t up to and including $t = 0$. When this is the case the simple transforms given in Table 1.1 apply; these may be verified from the definition 1.10 by integration by parts.

If, in any particular case, we take the Laplace transforms of both sides of the network linear differential equations (§ 1.2) we shall obtain a set of algebraic equations which may then be solved for the transforms of the voltages and currents in terms of the transforms of the e.m.fs. Let us instead apply the operation to the equations 1.1 and 1.2. From the list of transforms we see that equation 1.1 for an inductor yields

$$\bar{V} = pL\bar{I} \tag{1.11}$$

and for a capacitor relation 1.2 gives

$$\bar{V} = \bar{I}/pC \tag{1.12}$$

provided the capacitor is uncharged for $t \leq 0$.

KIRCHHOFF's laws and OHM's law for a resistance continue to apply to the new variables. We can therefore use the ordinary laws of fig. 1.1, and also THEVENIN's theorem, if we employ the Laplace transforms of the variables and the operational impedances pL, $1/pC$

TABLE 1.1. Some elementary Laplace transforms.

Function of time	Transform
$f(t) \quad f(t) = 0 \quad \text{for } t \leq 0$	$\bar{f}$
$\dfrac{df}{dt}$	$p\bar{f}$
$\dfrac{d^n f}{dt^n}$ all lower derivatives zero at $t = 0$ $\quad n = 0, 1, 2 \ldots$	$p^n \bar{f}$
$\displaystyle\int_0^t f(t_1) dt_1$	$\dfrac{1}{p} \bar{f}$
$\displaystyle\int_0^t \int_0^{t_1} \ldots \int_0^{t_{n-1}} f(t_n) dt_n \ldots dt_1 \quad n = 0, 1, 2 \ldots$	$\dfrac{1}{p^n} \bar{f}$
$[1]_0$	$\dfrac{1}{p}$
$\dfrac{t^n}{n!} \quad n = 0, 1, 2 \ldots$	$\dfrac{1}{p^{n+1}}$
$e^{-t/T}$	$\dfrac{T}{1 + pT}$
$1 - e^{-t/T}$	$\dfrac{1}{p(1 + pT)}$
$f(t - T)$	$e^{-pT} \bar{f}$
$te^{-\alpha t}$	$\dfrac{1}{(p + \alpha)^2}$
$e^{-\alpha t} \sinh \beta t$	$\dfrac{\beta}{p^2 + 2\alpha p + (\alpha^2 - \beta^2)}$

(and R). The final result relating the transform $\bar{I}$ of the output to that of the input $\bar{E}$, will be of the form

$$\bar{I} = G(p)\bar{E} \qquad (1.13)$$

where $G(p)$ is some function of p characteristic of the particular network. The input E may be any function of time but $\bar{E}$ is a function of p only; the output $\bar{I}$ is accordingly given entirely by a function of p.

The problem remains of how to find the inverse transformation so that the output may be obtained as a function of time. The Fourier-Mellin Inversion Integral, in the complex p-plane, gives the required result

$$f(t) = \frac{1}{2\pi j} \int_{c-j\infty}^{c+j\infty} e^{pt} \bar{f} dp \qquad (1.14)$$

where c is a positive constant chosen sufficiently large such that all the singularities of $\bar{f}$ lie to the left of the line of integration. The solution may often be obtained by reference to lists of standard forms to be found in the books by CARSLAW and JAEGER, GARDNER and BARNES (see Bibliography) and elsewhere.

1.5.2 Approximations.

The correspondence between the sinusoidal and Laplacian approaches is indicated by comparing the complex reactances $j\omega L$ and $1/j\omega C$ with the operational impedances pL and $1/pC$. It is quite generally true that the result for a sinusoidal input, of angular frequency ω, can be obtained from equation 1.13 simply by replacing p by $j\omega$ in the function $G(p)$; the quantities $\bar{E}$ and $\bar{I}$ then become the complex amplitudes.

The operation of taking the inverse Laplace transform is often difficult and it is very useful to fall back on frequency component concepts in order that approximations may be introduced. The correspondence shows that large values of p are associated with high frequency components that is with (i) short pulses, (ii) the leading edges of long sharp pulses, or (iii) with short time intervals. Accordingly when we are interested in what happens to the leading edge of a pulse we may obtain an approximation to part of the solution by neglecting all except the higher powers of p in the function $G(p)$. Alternatively, if we are concerned with the general treatment suffered, throughout its whole length, by a smooth more or less flat topped pulse low frequency components are relevant and only the lower powers of p need be taken into account.

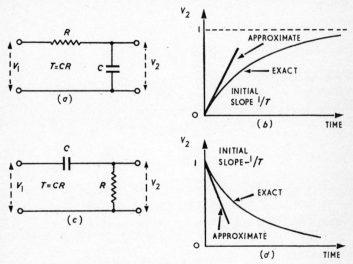

Fig. 1.5. (a) Simple integrating circuit.
(b) Approximate and exact reponse to unit step function input.
(c) Simple differentiating circuit.
(d) Response.

Let the circuit of fig. 1.5(a) be considered by way of example. Here

$$\bar{V}_2 = \frac{1}{1+pT} \cdot \bar{V}_1 \qquad 1.15)$$

At high frequencies, such that $\omega T \gg 1$, we may write

$$\bar{V}_2 \simeq \frac{1}{pT} \cdot \bar{V}_1 \qquad (1.16)$$

whence, on taking the inverse transform (see Table 1.1), we have

$$V_2(t) \simeq \frac{1}{T} \int_0^t V_1(t')dt' \qquad (1.17)$$

For short pulses, of duration much less than T, this simple arrangement therefore behaves as an integrating circuit. The effect on a unit step function, as predicted by equation 1.17, is compared with the familiar exact exponential solution in fig. 1.5(b).

At low frequencies, on the other hand, equation 1.15 may be written as
$$\bar{V}_2 \simeq (1 - pT) \cdot \bar{V}_1 \simeq e^{-pT} \cdot \bar{V}_1 \qquad (1.18)$$
i.e.
$$V_2(t) \simeq V_1(t - T) \qquad (1.19)$$

The circuit thus has the effect of a delay on pulses which change smoothly and by small fractional amounts in times of the order of T (a step function is therefore excluded).

Turning now to the arrangement of fig. 1.5(c) it is seen that
$$\bar{V}_2 = \frac{pT}{1 + pT} \cdot \bar{V}_1 \qquad (1.20)$$

When the pulses are short compared with T, i.e. $\omega T \gg 1$, we have $\bar{V}_2 \simeq \bar{V}_1$ and hence $V_2(t) \simeq V_1(t)$; the circuit to a first approximation, does not modify the pulse in any way.

As a second approximation, still at high frequencies, equation 1.20 may be written
$$\bar{V}_2 = \frac{1}{1 + 1/pT} \cdot \bar{V}_1 \simeq \left(1 - \frac{1}{pT}\right) \cdot \bar{V}_1 \qquad (1.21)$$

On taking the inverse transform we find
$$V_2(t) \simeq V_1(t) - \frac{1}{T} \int_0^t V_1(t') dt' \qquad (1.22)$$

The output pulse is thus equal to the input pulse but has subtracted from it a quantity proportional to the time integral of the input. The effect on a unit step function input, as determined by the approximate result (equation 1.22), is compared in fig. 1.5(d) with the well known exact exponential solution.

For pulses long compared with T, or for pulses which change smoothly and only by small fractional amounts in intervals of time T (thus excluding a step function), we have $\omega T \ll 1$ and
$$\bar{V}_2 \simeq pT(1 - pT) \cdot \bar{V}_1 \simeq pTe^{-pT} \cdot \bar{V}_1 \qquad (1.23)$$
whence
$$V_2(t) \simeq T \frac{dV_1(t - T)}{dt} \qquad (1.24)$$

The circuit then has a differentiating action coupled with a (small) delay.

1.6 SIMPLE VARIABLE CIRCUIT

The theory outlined above only applies, as has been pointed out, to linear invariable circuits among which may be listed transmission lines, delay lines, transformers, attenuators, amplifiers and the cathode ray oscillograph. We shall also have to deal with trigger circuits of various types and with discharge circuits involving relays and thyratrons. Such arrangements may be both non-linear and variable but usually they possess two stable states i.e. give rise to a simple switching action. When such is the case no difficulty arises; linear theory applies, with the help of the theorem below, provided we are not interested in what happens during the actual switch-over period.

Suppose we require to find the effect in any particular branch PQ of an otherwise linear invariable network subsequent to the closing of a switch, at time t_1, connected between any two points A and B of the network. The circuit may be reduced to an invariable one by imagining a voltage generator $E(t)$ (of zero impedance) to be connected permanently across AB. $E(t)$ is to be chosen such that the short-circuit has no effect on the network for $t < t_1$; and when $t > t_1$ we make $E(t) = 0$ thus allowing the influence of the short-circuit to be felt from t_1 onwards. The following procedure is adopted:

(i) In the absence of the short-circuit, find the voltage $V(t)$ appearing across AB for all times up to $t = t_1$ due to the internal e.m.fs. in the network.

(ii) Determine the response at PQ, due to the internal e.m.fs., for all times with the short-circuit present always.

(iii) Find the response at PQ for all t due to the e.m.f. $E(t) = V(t)$ for $t < t_1$ and $E(t) = 0$ for $t > t_1$ acting between A and B (the internal e.m.fs. may be supposed to be shut off for the purpose of this calculation).

The required response at PQ is then given by the sum of the responses (ii) and (iii). The salient contribution is due to the negative voltage step function $-[V(t_1)]_{t_1}$ which must be produced by the generator at time t_1.

1.7 CIRCUITS WITH DISTRIBUTED PARAMETERS

KIRCHHOFF's and OHM's laws apply, in the first instance, to circuits composed entirely of lumped elements. In such circuits the connecting wires play no part other than as a means of conveying current from one element to another. In any practical system the elements have a physical size and the properties are distributed in space; the connecting wires also play a further part. The latter, in particular, possess a stray capacitance to earth and also have self inductance; unwanted mutual capacitance and inductance may exist between various parts of the network. Components may assume a dual character; a coil possesses a certain self capacitance between the windings, for example, and a wire-wound resistor and certain types of capacitor are partly inductive. These effects become significant in the design of circuits for the microsecond range and all such parasitic or stray reactances are, of course, even more undesirable* in the millimicrosecond region. Being of subsidiary importance as far as the main circuit is concerned (though limiting performance at high frequencies), the strays may be allowed for in circuit analysis by the inclusion of small lumped series inductors and shunt capacitors connected at appropriate points in the network.

When ultra-high-frequency signal components are encountered stray reactances become a primary consideration. An obvious extension of lumped circuit theory to distributed circuit problems can be reached by allowing the number of circuit elements to increase and the effect of each to decrease correspondingly thus arriving, in the limit, at a continuous structure. The network equations give place to partial differential equations involving space as well as time as the independent variables. In most of the problems encountered in broad-band millimicrosecond pulse devices, as opposed to waveguide systems, such a treatment is adequate. The more general theory of electromagnetic waves must first be invoked, however, in order to justify this procedure. It is indicated in Appendix I that distributed circuit theory is rigorously applicable in the case of a smooth uniform transmission line operating in the principal mode.

---

* Great care must be taken in practice to avoid the use of components which suffer from these defects. At the high frequencies with which we are concerned, skin effect resistance in coils and dielectric loss in capacitors may also be important.

Each signal frequency component corresponds to a certain wavelength of electromagnetic radiation and, broadly speaking, one can say that lumped circuit approximations are adequate when the wavelength of the highest relevant frequency component is much greater than the spatial dimensions. The wavelength, for a given frequency, depends upon the wave velocity; normally the speed is that pertaining to a plane wave in the particular dielectric medium and this will be implied throughout unless otherwise stated. The velocity may, however, be very much less than this as in the important special case of waves on a long helix. When any one dimension is not very much less than the shortest relevant wavelength, distributed circuit theory may apply but electromagnetic theory must be taken as the ultimate basis in all cases of doubt.

The special interest attached to the millimicrosecond pulse region is evinced here. On the one hand, adopting the lumped equivalent of distributed parameters, we attempt to improve upon conventional microsecond techniques, and on the other, become concerned with devices which exhibit wave propagating properties and which depend for their operation on the essentially distributed nature of the circuit elements.

2

TRANSMISSION LINES

2.1 INTRODUCTION

THE problem of the distortionless transmission of signals over wires is a familiar one in electrical communication systems. In the millimicrosecond range the problem becomes acute, owing to the large bandwidths involved, and is met with even when the distances encountered are as small as a few inches. Difficulties arise in the layout and construction of electronic units due to the series self inductance and shunt capacitance to earth of every piece of connecting wire employed.

The theory of transmission lines, which has been well developed in the field of radio, indicates how such problems may be solved. We shall find that a pulse can be propagated down a line without distortion and that the input impedance of any length of a properly terminated line is a pure resistance. Upon these two properties hinge most applications of lines and we can see that all inter-connections must either be extremely short in length or must form a properly designed transmission line, which may then be of any length, working between the correct terminal impedances.

A survey of conventional transmission line theory, using the method of Laplace transforms, will be given here since lines find wide application in the equipments with which we are concerned and also form the basis of several devices which are peculiarly suited for use in the millimicrosecond range.

2.2 UNIFORM RECTILINEAR LINES

2.2.1 Summary of Properties.

Treatments based on MAXWELL's equations are given in the literature (see Bibliography) and some universal properties of the electric and magnetic wave fields are derived in Appendix I. The following is a summary of general results which (except (v)) apply to lines consisting of sensibly perfect conductors:

(i) A transmission line must consist of at least two* separate conductors. If these are parallel to one another and of constant, though arbitrary, cross-section along the line the system may be referred to as a rectilinear uniform line.

(ii) The solution of MAXWELL's equations, with the appropriate boundary conditions, corresponding to the simplest possible field pattern, shows the existence of a travelling wave (in either direction) in the principal or transverse electromagnetic mode. Properties possessed by waves in this mode are:

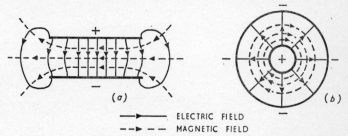

Fig. 2.1. Field distributions in transverse plane for (a) parallel strip and (b) coaxial transmission lines. (Wave travelling into plane of diagram.)

(iii) The electric and magnetic fields are always mutually perpendicular and at right-angles to the direction of propagation.

(iv) The characteristic impedance (see later) is a pure resistance and its value is independent of frequency.

(v) When losses are present (see § 2.2.6) the attenuation always increases with increasing frequency.

(vi) The electric and magnetic field distributions, in any transverse plane, are the same as those for electrostatic charges situated on the conductors and for direct currents† flowing through them respectively. Figs. 2.1(a) and (b) show the respective configurations in the case of a parallel strip transmission line and a circular coaxial line.

---

* For information on multiple conductor lines see articles by BOAST [201], BRAZMA [202, 203], BRÜDERLINK [204], FRANKEL [205], and KLUSS [206].

† Because of the skin effect the high frequency currents flow only in the surface layers of the conductors; the analogous direct currents must therefore be supposed to be confined to such layers, which, for perfect conductors, are infinitely thin.

(vii) It is possible to think in terms of a potential difference across the two conductors and equal and opposite currents flowing along them. A shunt capacitance C per unit length and a series inductance L per unit length can be introduced the values of which may be determined by static or quasi-stationary theory (see previous footnote).

(viii) The velocity of propagation is independent of frequency; in particular, there does not exist any low frequency cut-off as obtains in a waveguide. The speed is equal to that of a plane wave in an infinite volume of the dielectric which fills the space between the conductors. If T is the time delay per unit length then

$$T = \sqrt{\varepsilon\mu} = \frac{10}{3}\sqrt{\kappa_e \kappa_m} \text{ m}\mu \text{ sec./m} \qquad (2.1)$$

where $\kappa_e = \varepsilon/\varepsilon_0$ is the relative permittivity of the medium, and $\kappa_m = \mu/\mu_0$ is the relative permeability (the quantities ε_0 and μ_0 refer to free space).

(ix) We shall see that the characteristic impedance Z_0 is given by $\sqrt{L/C}$ and also that $T = \sqrt{LC}$ thus, by relation 2.1, if either L or C is known Z_0 may be determined. The calculation of characteristic impedances accordingly reduces to the determination of either L or C in the static case.

The application of distributed circuit theory is justified over regions where the line is uniform, or when the transverse dimensions

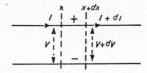

Fig. 2.2. Transmission line analysis.

change only slowly with distance along the line. The effect of discontinuities, however, must be investigated by means of electromagnetic theory. We shall now proceed to elaborate some of the above statements.

2.2.2 Analysis.

Turning to the conventional distributed circuit analysis of a transmission line (fig. 2.2), let $V_x(t)$ and $I_x(t)$ be the voltage and current at time t at a point x on the line. By considering the potential fall across a small distance dx due to the current flowing in the series inductance $L \cdot dx$ we have

$$\frac{d\bar{V}_x}{dx} = -pL \cdot \bar{I}_x \tag{2.2}$$

where the Laplace transforms,* with respect to time, of the voltage and current have been taken. Consideration of the drop in current over the element dx due to the voltage acting across the shunt capacitance $C \cdot dx$ shows that

$$\frac{d\bar{I}_x}{dx} = -pC \cdot \bar{V}_x \tag{2.3}$$

On differentiating both sides of equation 2.2 with respect to x and substituting for $d\bar{I}_x/dx$ from relation 2.3 we obtain

$$\frac{d^2\bar{V}_x}{dx^2} = p^2 LC \cdot \bar{V}_x \tag{2.4}$$

provided L is independent of x.

Similarly it follows that

$$\frac{d^2\bar{I}_x}{dx^2} = p^2 LC \cdot \bar{I}_x \tag{2.5}$$

provided C is independent of x.

It is easily verified that

$$\vec{\bar{V}}_x = \vec{\bar{V}}_0 e^{-pTx} = \vec{\bar{V}}_x \quad \text{say} \tag{2.6}$$

and

$$\overleftarrow{\bar{V}}_x = \overleftarrow{\bar{V}}_0 e^{pTx} = \overleftarrow{\bar{V}}_x \quad \text{say} \tag{2.7}$$

are each solutions of equation 2.4 where $\vec{\bar{V}}_0$ and $\overleftarrow{\bar{V}}_0$ are constants, provided both L and C are independent of x (the significance of the arrows will appear in a moment). We have put

$$T = \sqrt{LC} \tag{2.8}$$

---

* A Laplace transform treatment of transmission lines has been given by WAIDELICH [207].

TRANSMISSION LINES

The general solution of equation 2.4 is the sum of the solutions 2.6 and 2.7 and accordingly

$$\bar{V}_x = \overrightarrow{\bar{V}}_x + \overleftarrow{\bar{V}}_x = \overrightarrow{\bar{V}}_0 e^{-pTx} + \overleftarrow{\bar{V}}_0 e^{pTx} \qquad (2.9)$$

On taking the inverse Laplace transform (see Table 1.1, page 10) this becomes

$$V_x(t) = \overrightarrow{V}_x(t) + \overleftarrow{V}_x(t) = \overrightarrow{V}_0(t - Tx) + \overleftarrow{V}_0(t + Tx) \qquad (2.10)$$

provided $t - Tx > 0$.

The first term on the right hand side of the solution 2.10 shows that the voltage at the point x at time $t + Tx$ is the same as that at $x = 0$ at time t i.e. we have propagation, without distortion, with delay time T per unit length. The proviso $t - Tx > 0$ indicates that we must not use this term to give the effect at x until the wave has had time to travel from $x = 0$. Similarly the second term represents a wave travelling with the same speed but along the negative direction of x.

Exactly the same arguments apply for the solution of the current wave equation 2.5. We have only to introduce two new constants and in place of equations 2.6, 2.7 and 2.9 we have

$$\bar{I}_x = \overrightarrow{\bar{I}}_x + \overleftarrow{\bar{I}}_x \qquad (2.11)$$

where

$$\overrightarrow{\bar{I}}_x = \overrightarrow{\bar{I}}_0 e^{-pTx} \qquad (2.12)$$

and

$$\overleftarrow{\bar{I}}_x = \overleftarrow{\bar{I}}_0 e^{pTx} \qquad (2.13)$$

The complete solutions 2.9 and 2.11 must satisfy equation 2.2 (or 2.3) for all values of x. If the quantity Z_0 is defined by

$$Z_0 = \sqrt{L/C} \qquad (2.14)$$

it is easily shown that

$$\frac{\overrightarrow{\bar{V}}_0}{\overrightarrow{\bar{I}}_0} = \frac{\overrightarrow{\bar{V}}_x}{\overrightarrow{\bar{I}}_x} = -\frac{\overleftarrow{\bar{V}}_0}{\overleftarrow{\bar{I}}_0} = -\frac{\overleftarrow{\bar{V}}_x}{\overleftarrow{\bar{I}}_x} = Z_0 \qquad (2.15)$$

Since Z_0 does not involve p the operation of taking the inverse transform throughout relation 2.15 consists simply of the removal of the bars. We therefore see that the characteristic impedance Z_0 (which is purely resistive and independent of frequency) gives the ratio of the voltage to the current in the forward going wave component; the ratio is the same for the backward travelling wave but with the sign changed.

2.2.3 General.

Fig. 2.1 shows only two of many possible forms of transmission line; papers by ANDERSON [208], BARCLAY and SPANGENBERG [209], BROWN [210], BUCHHOLZ [211], CRAGGS and TRANTER [212], FRANKEL [213, 214], GANS [215], LANDSBERG [216], MEINKE [217], PARZEN [218], RE QUA [219], ROTHE [220], SHEBES [221], TSEITLIN [222, 223, 224], WHEELER [225] and WISE [226], contain information on lines having various cross-sectional shapes (including curvilinear lines). The characteristic impedances of a number of configurations are given in Appendix II.

In the majority of applications we are concerned with an unbalanced system, that is, one in which voltages with respect to earth are relevant. When such is the case a single insulated conductor surrounded completely by a metal shield may be employed. The shield acts both as the second conducting element and as a screen. Owing to the skin effect, the current in the shield flows entirely on its inner surface and the outside of the shield is completely "dead" and may be earthed anywhere or everywhere without affecting operation (but see § 3.5). For details of the design, construction and properties of coaxial cables the reader may consult papers by KENNEY [227], SMITH [228], STANFORD and QUARMBY [229], ZIMMERMAN [230] and the Conference Report [231].

When the system is balanced, two insulated conductors may be employed, again completely surrounded by a shield. Totally screened lines are preferable to open types since the latter are subject to a frequency dependent loss of energy by radiation and unwanted signals may be picked up in nearby circuits (or vice-versa).

2.2.4 Terminations and Discontinuities

We have been concerned in the previous sections with the properties of the main body of the line and now investigate the effect of discontinuities which may be present.

2.2.4.1 Arbitrary Termination

The most important, and most obvious, discontinuities occur at the two ends of a finite length of line. Let us suppose that an arbitrary impedance Z, a function of p, is connected across the right hand, or far, end of the line i.e. the end remote from the

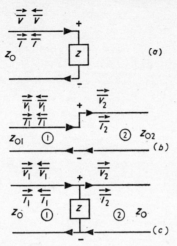

Fig. 2.3. Major discontinuities in a transmission line.
(a) Abitrary termination.
(b) Junction of two lines of different characteristic impedances.
(c) Shunt impedance connected across a uniform line.

generator feeding the line. If the amplitudes of the waves, just at the termination, are as indicated in fig. 2.3(a) OHM's law gives

$$Z = \frac{\vec{V} + \overleftarrow{V}}{\vec{I} + \overleftarrow{I}} \qquad (2.16)$$

Substituting for the currents in terms of the voltages by means of relations 2.15 we find for the voltage reflection coefficient

$$\rho \equiv \frac{\overleftarrow{V}}{\vec{V}} = \frac{Z - Z_0}{Z + Z_0} \qquad (2.17)$$

Three particular cases are of special interest:

(a) If $Z = Z_0$ then $\rho = 0$. There is no reflected wave and all the power is absorbed in the pure resistance Z. The line is said to be matched or properly terminated.

(b) If $Z = 0$, corresponding to a short circuit, then $\rho = -1$ and the reflected voltage wave is of the same amplitude as the incident one but opposite in sign.

(c) If $Z \to \infty$, as is the case when the line is open-circuited, $\rho = 1$ and reflection occurs without change in phase.

We may at once apply relation 2.17 to the case of the junction between two lines of different characteristic impedances (fig. 2.3(b)). No loss in generality is incurred if we suppose that there is only one incident wave i.e. that travelling to the right in line 1. Relations 2.15 show that the input impedance $\vec{V}_2/\vec{I}_2$ of the line 2 is simply Z_{02} and this quantity is accordingly substituted for Z in the previous equation yielding

$$\rho = \frac{Z_{02} - Z_{01}}{Z_{02} + Z_{01}} \qquad (2.18)$$

In the absence of any extra lumped series impedance at the junction we must have

$$\vec{V}_1 + \overleftarrow{V}_1 = \vec{V}_2 \qquad (2.19)$$

and accordingly a voltage transmission coefficient can be defined as

$$\frac{\vec{V}_2}{\vec{V}_1} = 1 + \frac{\overleftarrow{V}_1}{\vec{V}_1} = 1 + \rho \qquad (2.20)$$

When $Z_{01} = Z_{02}$ then $\rho = 0$ and the transmission coefficient equals unity; no reflection occurs at the junction of two lines of the same impedance and the transmission is complete.

Let us now consider the case of an arbitrary impedance Z connected across a uniform line at some intermediate point (fig. 2.3(c)). The effective impedance terminating the left hand section of line is equal to $ZZ_0/(Z + Z_0)$ and equation 2.17 gives

$$\rho = \frac{-Z_0}{2Z + Z_0} \qquad (2.21)$$

In this case the voltage is continuous over the discontinuity and relations 2.19 and 2.20 are again satisfied.

Similar arguments may be applied in the determination of current reflection and transmission coefficients for the cases indicated in fig. 2.3. In case (c) of this figure the current is not continuous, however, over the discontinuity and the equation in current analogous to 2.19 is not true (similarly relation 2.19 for the voltage would not obtain if a lumped series impedance were inserted in the line). The coefficients are found in all cases by employing relations 2.15 in each section of line on either side of the discontinuity and applying OHM's law and the appropriate continuity relation.

The presence of a discontinuity on the line with its attendant reflection represents an effective attenuation of the main wave. If the discontinuity is brought about by purely resistive impedances, as in fig. 2.3(b) or in fig. 2.3(c) when Z is a resistance, the reflection coefficient is independent of p and of frequency; no distortion occurs. When the discontinuity involves reactances ρ is a function of p and the "attenuation" is accompanied by distortion.

In the realm of sinusoidal oscillations reflection coefficients are usually measured in terms of the standing wave ratio S. The following relation obtains

$$S = \frac{|\vec{V}| + |\overleftarrow{V}|}{|\vec{V}| - |\overleftarrow{V}|} = \frac{1 + |\rho|}{1 - |\rho|} \qquad (2.22)$$

where $|\vec{V}|$ and $|\overleftarrow{V}|$ are the ordinary amplitudes of the waves travelling in the two directions. When $|\rho|$ is small this relation gives $|\rho| \simeq (S - 1)/2$.

The above treatment of discontinuities by means of circuit theory, represents an oversimplification. We shall pass on to consider further effects which may become of importance at the high frequency end of the millimicrosecond range.

2.2.4.2 Higher Modes

In addition to the principal mode, corresponding to the simplest field pattern, an infinite variety of field configurations can be

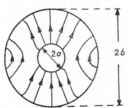

Fig. 2.4. Electric field pattern in a coaxial line for the next mode above the principal.

found (see BONDI and KUHN [232]) which satisfy MAXWELL's equations between the conductors and the boundary conditions upon them. The field pattern, in the case of the coaxial line, for the next mode above the principal is shown in fig. 2.4. The general properties of these higher modes are as follows:

(i) The electric and magnetic fields are not perpendicular to the direction of propagation.

(ii) The phase velocity of propagation depends on the frequency.

(iii) There exists a certain cut-off frequency for each mode such that there is no real transmission of energy down the line at lower frequencies; the fields die away and become negligible at a distance down the line of the order of its diameter.

The H_{11} or TE_{11} mode illustrated has the lowest cut-off frequency f_c and the corresponding plane wavelength λ_c is given approximately by

$$\lambda_c = \frac{c}{f_c \sqrt{\kappa_e \kappa_m}} \simeq \pi(a + b)$$

where c is the velocity of light.

Consider the discontinuity indicated in fig. 2.5(a). It is impossible to match the fields on the two sides of the discontinuity simply by

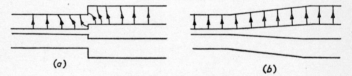

Fig. 2.5. (a) Field pattern in coaxial line involving higher modes at a discontinuity.
(b) Smooth transition from one set of transverse dimensions to another without change in characteristic impedance.

combining waves in the principal mode. Higher modes are called into play in order that the boundary conditions and field continuity conditions may be satisfied everywhere. At frequencies below their respective cut-offs the higher modes cannot transmit energy and consequently part of the incident wave is reflected. The effect is more pronounced at the higher frequencies and the high frequency components in the incident signal will be partially reflected with a consequent distortion of the main wave.

Reflection occurs even if the characteristic impedance of the line is the same on both sides of the discontinuity. When it is desired to effect a change of dimensions, at constant characteristic impedance, the change should be made smoothly. If the length of the tapered portion (fig. 2.5(b)) is several times the diameter of the outer conductor, and the ratio of the outer to inner diameters is maintained constant, the fields will not sensibly depart from the principal mode type (a pair of coaxial cones can support a principal wave at constant impedance). There will accordingly be negligible reflection.

WHINNERY *et al.* [233, 234] and MILES [235] have shown how the effect of various types of discontinuity may be represented, for

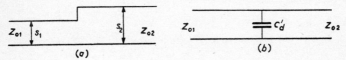

Fig. 2.6. (a) Discontinuity in parallel strip transmission line.
(b) Equivalent circuit.

principal mode circuit theory to apply, by a shunt capacitance connected at the junction. Thus, for example, the parallel plate line of fig. 2.6(a) is equivalent to that shown schematically in fig. 2.6(b) where the capacitance C'_d is given by the following Table.

TABLE 2.1

s_1/s_2	C'_d (pF/cm width)
0·0	∞
0·2	0·070
0·4	0·033
0·6	0·015
0·8	0·004
1·0	0·000

If a uniform dielectric is present these values for the capacitance must be multiplied by κ_e. When dielectric material is to be found

Fig. 2.7. Effect of dielectric on one side of a discontinuity.
C'_d is multiplied by κ_e in (a) but not in (b).

only on one side of the boundary it is taken into account if it lies on the side where the field disturbance is the greater. Thus, in fig. 2.7(a) the value of C'_d is multiplied by κ_e but is left unaltered in the case of fig. 2.7(b).

For a coaxial line the equivalent shunt capacitance C_d is given by

$$C_d = 2\pi r C'_d$$

where C'_d is the value tabulated above, per unit width, for parallel plates. The representative radius r for a number of cases is quoted in fig. 2.8.

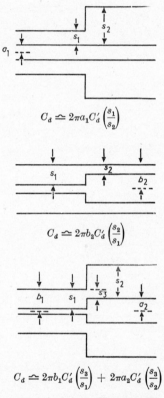

Fig. 2.8. Capacitances equivalent to discontinuities in coaxial line.

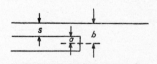

Fig. 2.9. Transmission line with break in inner conductor.

An open circuit termination (fig. 2.9) is equivalent to an ideal line terminated by a capacitance of value given below.

TABLE 2.2

a/b	$C_d/2\pi b$ (pF/cm)
0·0	0·000
0·2	0·016
0·4	0·034
0·6	0·056
0·8	0·095
1·0	∞

The equivalent capacitance concept must be applied with the following reservations:

(i) The results are limited to frequencies where the plane wavelength is somewhat greater than the largest transverse dimension. When the frequency approaches the cut-off value for one of the higher modes the value of C_d ceases to be frequency independent.

(ii) If multiple discontinuities are present close to one another mutual interaction occurs and the equivalent capacitances are modified (see MARCUS [236]).

(iii) Similarly the results must be corrected if the line is terminated near a discontinuity ("near" implying within a distance of the order of the greatest transverse dimension).

The distributed circuit methods of § 2.2.4.1 only apply exactly to cases in which the fields at the discontinuity can be accurately matched in terms of waves in the principal mode. The only two cases are:

(i) A change in dielectric material, in what would otherwise be a uniform line, provided the boundary lies in the transverse plane.

(ii) A short-circuit termination, provided this consists of a flat plate or disc placed perpendicularly across the end of the line.

2.2.4.3 Supports

Supports of dielectric material are sometimes employed in the construction of coaxial cable. If these are equally spaced strong reflections can occur at frequencies for which the supports are an integral number of half-wavelengths apart. Such types should be avoided unless the supports are very close together.

Supports are often unavoidable and CORNES [237] has shown how the reflection coefficient may be made negligible. As illustrated in fig. 2.10(a), the diameter of the inner conductor may be reduced by an amount sufficient to compensate for the influence of the dielectric

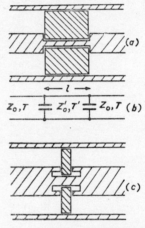

Fig. 2.10. (a) Dielectric support.
(b) Circuit equivalent to (a).
(c) Compensated disc support.

on the characteristic impedance. This is perfectly satisfactory at medium frequencies but at the higher frequencies the effect of the discontinuity in radius becomes important especially if many supports are employed. If l is the thickness of the support, the equivalent circuit is as shown in fig. 2.10(b) where Z_0' is the characteristic impedance of the short length of line constituted by the support. Clearly l should be as small as possible and the dielectric constant of the insulator should approach unity. Circuit analysis shows that Z_0' should be about 8% greater than Z_0 in a typical case in order to offset the effect of the capacitances. The voltage reflection coefficient is found to be less than 1% for all frequencies up to 4000 Mc/s.

KADEN and ELLENBERGER [238] and PETERSON [239] recommend a support with concave faces and describe a modification in which the unwanted shunt capacitance is compensated by the inclusion of a series inductance. This is achieved by undercutting the inner conductor* in the manner indicated in fig. 2.10(c). The extent of the undercut was determined experimentally and a reflection coefficient of less than 0·4% was obtained up to a frequency of 1200 Mc/s. In a line containing 20 supports, spaced half a wavelength apart, the voltage standing wave ratio was reduced from 2·1 to 1·05 by the introduction of the compensation.

It is evident from the foregoing that the number of plug and socket joints in a run of cable should be kept to a minimum and those that are unavoidable should be designed to give a good match over a wide band of frequencies. The junction will consist of short lengths of line, of differing impedances, together with the shunt capacitances C_d representing the effect of abrupt changes in dimensions. If the wavelength is much longer than the linear dimensions of the junction (as will be the case in the millimicrosecond region) then each section of length l_n, inductance per unit length L_n, and capacitance per unit length C_n, may be treated as a lumped circuit possessing series inductance $l_n L_n$ and shunt capacitance $l_n C_n + C_{dn}$.

When a short length of line, of characteristic impedance Z_0', and total delay time T (single transit), is inserted in a line of impedance Z_0, it may be shown that a voltage pulse of amplitude $2T(Z_0' - Z_0)/t_r(Z_0' + Z_0)$ is reflected when the input pulse is of unit amplitude and has a rise-time t_r. This formula, which obtains when $2T < t_r$ and $Z_0' \sim Z_0$, may be applied in the case of a connector. A similar result is quoted by Garwin [241].

2.2.4.4 Terminal Matching

We have seen that ideally no reflection occurs when a line is terminated by a resistance equal in value to its characteristic impedance. The connection of a lumped resistor across the line introduces stray reactances which spoil the match at high frequencies; an improvement can be obtained by attempting to satisfy the field boundary conditions by employing a disc type resistor. CROSBY and PENNYPACKER [242] have studied the behaviour of resistors of the type where the resistive element is in the form of a film deposited on an insulating rod.† The resistor is housed in a cylindrical container, of length much greater than its diameter, short-circuited at

---

* Such an undercut inner conductor may be useful for other matching purposes and is discussed further by ROWLAND [240]. (The slightly higher value of impedance Z_0' used by CORNES gives a compensatory series inductance; see § 2.2.5.3.)

† Further details of resistive components are given in the section on attenuators § 4.5.

one end. The lossy transmission line so formed (see § 2.2.6) is analysed and the input impedance determined. The total resistance R should be equal to the characteristic impedance of the line to be matched and it is found that optimum performance is obtained when $R/Z_0' = \sqrt{3}$ where Z_0' is the nominal characteristic impedance of the lossy section. A voltage reflection coefficient of about 1% is then found when the ratio of length to wavelength is 0·05; performance improves as this ratio decreases. The length must, however, remain considerably greater than the diameter in order that principal mode theory, which has been used, may apply.

CLEMENS [243] treats a similar case and gives the theory of a lossy exponential line; the arrangements are intended primarily for use in the microwave region however.

2.2.4.5 Other Types of Discontinuity

Further miscellaneous information on the effects of irregularities and discontinuities is given in papers by COUANAULT and HERRENG [244], COX [245], FUCHS [246], HERRENG and VILLE [247], RAYMOND [248] and SAPIR [249].

KING and TOMIYASU [250] give an account of the effect of the geometry of the connection to the load. In practice, terminal discontinuities are equivalent to a shunt capacitance connected across the line together with an inductance in series with the idealized lumped load.

TOMIYASU [251] finds that the effect of a sharp bend in a transmission line is equivalent to the insertion of a shunt capacitance, followed by a series inductance, which in turn is succeeded by a second shunt capacitance. OLIVER [252] gives an account of the influence of discontinuities in slotted concentric line impedance measuring gear over the range 100–3000 Mc/s.

MARCHAND [253, 254] enumerates a number of theorems concerning breaks in the outer shield of a screened line. At frequencies greater than 50 Mc/s the shield can be assumed to be perfect in the sense that the current on the inside surface and that flowing on the outside surface (if any) are quite independent of each other. KIRCHHOFF's laws may then be applied at the break in order to relate the currents.

2.2.5 Transmission line as a Circuit Element

Relations 2.1, 2.6, 2.7, 2.15 and 2.17 together with the discontinuity equivalents given above tell us all we need to know concerning the properties of a transmission line as a circuit element. In the applications to be described we shall see how the three basic properties which a line possesses namely (i) time delay, (ii) resistive characteristic impedance and (iii) reflection phenomena, are turned to advantage. The extensive use of lines in millimicrosecond equipment is due to the fact that the lengths involved are not inconveniently great as would usually be the case in the microsecond region.

2.2.5.1 *Expression for Voltage at Input End*

Fig. 2.11 depicts a source of e.m.f. $\bar{E}$, of internal impedance Z_1, situated at $x = 0$, feeding a length l of line, of characteristic im-

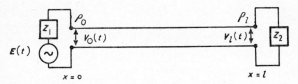

Fig. 2.11. General case for analysis.

pedance Z_0, which is terminated at the point $x = l$ by an arbitrary impedance Z_2. We require to find the voltage on the line at the input end.

From equation 2.9 the total voltage $\bar{V}_0$ at $x = 0$ consists of the sum of two wave components $\vec{V}_0 + \overleftarrow{V}_0$. The former is made up of (i) a contribution due to the applied e.m.f. of amount $\bar{E} \cdot Z_0/(Z_1 + Z_0)$ given by the elementary potential divider formula and (ii) the fraction $\rho_0 \cdot \overleftarrow{V}_0$ of the wave incident from the right which is reflected at $x = 0$. Here ρ_0 is the voltage reflection coefficient at the near end for waves incident from the right. Thus

$$\vec{V}_0 = a \cdot \bar{E} + \rho_0 \cdot \overleftarrow{V}_0 \qquad (2.23)$$

where we have put

$$a \equiv \frac{Z_0}{Z_1 + Z_0} \qquad (2.24)$$

the straightforward attenuation factor.

From relations 2.6, 2.7 and 2.17 we have

$$\frac{\overleftarrow{V}_0}{\vec{V}_0} = \rho_l e^{-2pTl} \qquad (2.25)$$

On solving for $\vec{V}_0$ and $\overleftarrow{V}_0$ from relations 2.23 and 2.25 we obtain

$$\vec{V}_0 = a \cdot \frac{1}{1 - \rho_0 \rho_l e^{-2pTl}} \cdot \bar{E} \qquad (2.26)$$

and

$$\overleftarrow{V}_0 = a \cdot \frac{\rho_l e^{-2pTl}}{1 - \rho_0 \rho_l e^{-2pTl}} \cdot \bar{E} \qquad (2.27)$$

hence

$$\bar{V}_0 = \vec{V}_0 + \overleftarrow{V}_0 = a \cdot \frac{1 + \rho_l e^{-2pTl}}{1 - \rho_0 \rho_l e^{-2pTl}} \cdot \bar{E} \qquad (2.28)$$

This general result can be interpreted in familiar terms if the denominator is expanded* as a power series; thus

$$\bar{V}_0 = a(1 + \rho_l e^{-2pTl})[1 + \rho_0 \rho_l e^{-2pTl} + (\rho_0 \rho_l)^2 e^{-4pTl} + \ldots] \cdot \bar{E} \qquad (2.29)$$

When the inverse transform is taken, $V_0(t)$ is seen to consist of the voltage $a \cdot E(t)$ up to time $t = 2Tl$ followed by an added contribution $a\rho_l(1 + \rho_0) \cdot E(t - 2Tl)$ for $2Tl \leq t < 4Tl$. At time $t = 4Tl$ a further contribution of $a\rho_0 \rho_l^2 (1 + \rho_0) \cdot E(t - 4Tl)$ is added and so on. The extra contribution which appears when $2nTl \leq t \leq 2(n+1)Tl$ is equal to $a(\rho_0 \rho_l)^{n-1} \rho_l (1 + \rho_0) \cdot E(t - 2nTl)$ where n is an integer.

Relations 2.28 and 2.29 are quite general but in the interpretation just given we have supposed that both ρ_0 and ρ_l are independent of p (i.e. both Z_1 and Z_2 are pure resistances) when taking the inverse transform.

2.2.5.2 *Pulse Shaping*

Let us restrict ourselves to the case in which the line is matched at the source end. Then $Z_1 = Z_0$, $\rho_0 = 0$ and $a = 1/2$. Equation 2.28 then becomes

$$\bar{V}_0 = \tfrac{1}{2}(1 + \rho_l e^{-2pTl}) \cdot \bar{E} \qquad (2.30)$$

No multiple reflection takes place and there is only one extra contribution to $V_0(t)$ which occurs at and after time $t = 2Tl$. Three particular cases, already mentioned in § 2.2.4.1, are of importance:

(a) $Z_2 = Z_0$. Here $\rho_l = 0$ and $V_0(t) = E(t)/2$

(b) $Z_2 = 0$. In this case $\rho_l = -1$ and

$$V_0(t) = \tfrac{1}{2}[E(t) - E(t - 2Tl)]$$

---

* This expansion is rigorously justified since p is a positive quantity and the reflection coefficients are each less than or equal to unity.

This arrangement forms a very useful clipping circuit, which is superior to the differentiating circuit of fig. 1.5(b), since the portion of the pulse between $t = 0$ and $t = 2Tl$ is reproduced without any modification in shape.

(c) $Z_2 \to \infty$. Now $\rho_l = 1$ and accordingly relation 2.30 yields

$$V_0(t) = \tfrac{1}{2}[E(t) + E(t - 2Tl)]$$

The results for a unit step function input and for a ramp function input of rate of rise m are shown in fig. 2.12(a), (b). The waveforms

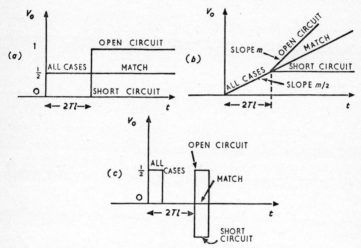

Fig. 2.12. Voltage at input of line with various types of termination.
    (a) unit step function input
    (b) ramp function input, slope m
    (c) short rectangular pulse input
    Source matched in all cases.

obtained with a short rectangular pulse of duration less than $2Tl$ (which are, of course, derived from the results for a step function) are depicted in fig. 2.12(c). It is seen that two pulses have been obtained from one with a relative delay between them of twice the transit time down the line. Such an artifice may be applied, for example, in a single channel double pulse generator unit.

When either the source or the terminating impedance is partially or wholly reactive the reflection coefficients involve p and this must be borne in mind when taking the inverse transform of equation 2.28. The results for a number of cases are summarized in fig. 2.13.

2.2.5.3 Input Impedance

The expression for the input impedance Z_{in} of a finite length of line, arbitrarily terminated, follows from the results of the preceding section. By the elementary potential divider formula

$$\bar{V}_0 = \frac{Z_{in}}{Z_{in} + Z_1} \cdot \bar{E} \tag{2.31}$$

which, on using relations 2.28, 2.24 and 2.17, yields

$$Z_{in} = Z_0 \frac{1 + \rho_l e^{-2pTl}}{1 - \rho_l e^{-2pTl}} \tag{2.32}$$

This result turns out to be independent of Z_1 as should indeed be the case if the input impedance is to have a definite significance. It is noted immediately that, if the line is correctly terminated at the far end ($\rho_l = 0$), then $Z_{in} = Z_0$ as expected.

We shall restrict ourselves to one special case, out of many of interest, and find the lumped circuit to which an arbitrarily terminated short length of line is equivalent. If $\omega Tl \ll 1$ we can write $e^{-2pTl} \simeq (1 - 2pTl)$ and equation 2.32 reduces to

$$Z_{in} \simeq Z_2 \frac{1 - pTl(Z_2 - Z_0)/Z_2}{1 + pTl(Z_2 - Z_0)/Z_0} \tag{2.33}$$

where we have substituted for ρ_l in terms of Z_2 and Z_0.

(a) If $Z_0 = Z_2$ then $Z_{in} = Z_2$.

(b) When $Z_0 \gg Z_2$ relation 2.33 gives

$$Z_{in} \simeq Z_2(1 + pTlZ_0/Z_2)$$
$$= Z_2 + pLl$$

on using equations 2.8 and 2.14. The effect of the line is to add a series lumped inductance equal to its total natural inductance.

(c) In the case $Z_0 \ll Z_2$ we find

$$Z_{in} \simeq \frac{Z_2}{1 + pTl/Z_0}$$

which is the expression for the combination of the impedance Z_2 in parallel with the total natural capacitance Cl of the line.

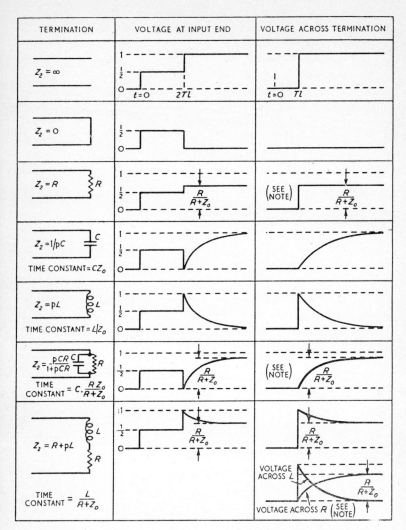

Fig. 2.13. Circuit properties of a length l of lossless uniform transmission line of characteristic impedance Z_0 when fed from a matched source giving a unit step function of e.m.f. at time $t = 0$.

NOTE: In these cases when R represents the input impedance of a second length of transmission line the signal launched in this line is of identical form.

We are also interested in the general expression for the voltage $\bar{V}_l$ which appears across the termination.
Now
$$\bar{V}_l = \vec{\bar{V}}_l + \overleftarrow{\bar{V}}_l = (1 + \rho_l)e^{-pTl} \cdot \vec{\bar{V}}_0 \qquad (2.34)$$
On applying relation 2.26 this becomes
$$\bar{V}_l = a\frac{(1+\rho_l)e^{-pTl}}{1-\rho_0\rho_l e^{-2pTl}} \cdot \bar{E} \qquad (2.35)$$
which, by virtue of equation 2.28, may be written in terms of the resultant voltage at the input
$$\bar{V}_l = \frac{(1+\rho_l)e^{-pTl}}{1+\rho_l e^{-2pTl}} \cdot \bar{V}_0 \qquad (2.36)$$

2.2.6 Losses.

Throughout the previous sections we have assumed that the transmission line was completely lossless. Losses may arise in practice due to:

(a) The finite resistance of the conductors.

(b) The conductivity of the dielectric (usually utterly negligible) and the radio-frequency losses therein,

(c) Radiation in the case of open, unscreened, lines.

It is usually assumed that if the losses are small the field patterns do not depart appreciably from those appertaining to the principal mode, and it turns out that all losses are equivalent to a distributed series resistance R and shunt conductance G per unit length (see BUCHHOLTZ [211] for example).

2.2.6.1 Effect on Propagation and Impedance

The basic equations 2.2 and 2.3 become
$$\frac{d\bar{V}_x}{dx} = -(R+pL)\bar{I}_x \qquad (2.37)$$

$$\frac{d\bar{I}_x}{dx} = -(G+pC)\bar{V}_x \qquad (2.38)$$

We note that all our previous relations involving the Laplace transforms of the variables still stand provided L is replaced by $L(1+R/pL)$ and C by $C(1+G/pC)$ throughout. If the frequency is high enough, as will normally be the case in the millimicrosecond range, such that $R \ll \omega L$ and $G \ll \omega C$ relations 2.8 and 2.14 become

$$T \simeq \sqrt{LC}\left[1 + \left(\frac{R}{L} + \frac{G}{C}\right)\bigg/2p\right] \tag{2.39}$$

$$Z_0 \simeq \sqrt{\frac{L}{C}}\left[1 + \left(\frac{R}{L} - \frac{G}{C}\right)\bigg/2p\right] \tag{2.40}$$

The solution 2.6 for a wave travelling to the right now reads

$$\vec{V}_x = \vec{V}_0 e^{-pTx} = \vec{V}_0 e^{-\alpha x} e^{-p\sqrt{LC}x} \tag{2.41}$$

where

$$\alpha \equiv \frac{R}{2}\sqrt{\frac{C}{L}} + \frac{G}{2}\sqrt{\frac{L}{C}} \tag{2.42}$$

We see that the delay time is unaffected, to the first order of terms in $1/p$, and an attenuating factor $e^{-\alpha x}$ has been introduced. The characteristic impedance has a value which differs from the old by a term in $1/p$.

2.2.6.2 Sources of Loss

When MAXWELL's equations are solved for a wave propagated in and over the surface of a metal of finite conductivity it is found that the field intensity falls off exponentially with distance below the surface. The current density is reduced to a fraction $1/e$ of its surface value at the skin depth d metres given by

$$d = \frac{1}{\sqrt{\pi f \mu \sigma}} \tag{2.43}$$

where f is the frequency* in c/s, σ is the conductivity in mhos/m and $\mu = \kappa_m \mu_0$ is the permeability of the conductor. At a frequency of 1 Mc/s, for example, with a copper conductor, $d \simeq 6 \times 10^{-3}$ cm and thus, in the millimicrosecond range, the currents may be assumed to flow entirely over the surface of the conductors. For the coaxial line, with copper conductors radii a and b, we have

$$R = 4 \cdot 2 \times 10^{-8}\sqrt{f}\left(\frac{1}{a} + \frac{1}{b}\right) \Omega/\text{m} \tag{2.44}$$

---

* Treatments of the skin effect for pulse signals are given by MILLER [255], SIMM [256, 257], and VALLESE [258].

The effective resistance increases with frequency but the ratio $R/\omega L$ continues to decrease as the frequency is raised thus justifying the approximations* 2.39 and 2.40. The attenuation is seen to rise with frequency resulting in a loss of the higher signal component frequencies. The distortion only appears as a slowing in the risetime, however, since there is no accompanying phase distortion (to the first order). A comprehensive collection of formulae and graphs for the skin effect in conductors of various shapes composed of different materials is given by WHINNERY [259].

It may be noted that skin resistance may be reduced by plating the conductors with silver to a thickness equal to several times the skin depth. It may also be mentioned that the resistance of an outer conductor (shield) composed of thin wires braided together may be several times that of the corresponding metal tube and the variation with frequency seems to be greater, at high frequencies, than predicted by simple skin effect theory.

It is possible to choose the line dimensions such that the attenuation is a minimum. For a coaxial line the attenuation constant has a flat minimum when $b/a = 3.6$ i.e. when $Z_0 = 77\ \Omega$. This and other optimum arrangements are discussed by SMITH [260], but such considerations are irrelevant except when long distances or high powers are involved.

The principal contribution to the shunt conductance G is that of the ordinary dielectric loss in the material given by

$$G = \omega C \tan \delta \qquad (2.45)$$

where $\tan \delta$ is approximately equal to the power factor when the losses are small. A list of permittivities and loss factors for various materials is quoted by MORENO [261], for example, and the relative merits of certain substances are discussed by WILLIAMS and SCHATZ [262]; BRECKINRIDGE and THURNAUER [263] have compiled a digest of the literature on dielectrics.

The attenuation found in practice, taking all sources of loss into account, is indicated in Table 2.3 for a number of different varieties of coaxial cable.

The Uniradio types 39 and 60 have solid polythene insulation and the T.C.M. type AS50 and Transradio C3-T are semi air-spaced. All these cables are mechanically flexible and long lengths can be stored

---

* The variation of R and G with frequency makes it impossible to obtain a rigorous circuit solution by Laplace transform methods.

conveniently. Uniradio 132, on the other hand, has a solid copper outer conductor and is filled with a mineral dielectric—hence the low attenuation found.

TABLE 2.3. Data for typical coaxial cables. Length for 100 mμsec. delay $\sim$ 80 ft.

Type		Characteristic Impedance Z_0 ohms	Overall diameter inches	Attenuation in db for 100 mμsec. delay				
British	American approximate equivalent			10 Mc/s	30 Mc/s	100 Mc/s	300 Mc/s	1000 Mc/s
Uniradio 39	RG/39U	69	0.31	0.51	0.85	1.65	3.3	7.8
Uniradio 60	RG/13U	75	0.41	0.41	0.65	1.2	2.3	5.0
T.C.M.–As50	RG/62U	100	0.28	0.58	1.0	1.9	3.3	7.0
Transradio —C3–T	—	197	0.64	0.43	0.85	1.6	3.6	10.2
Uniradio 132	—	42	0.63	0.06	0.14	0.19	0.36	0.78

Painting the conductors with a protective coating, such as lacquer, tends to introduce dielectric loss but the protection of the metal surface from oxidation and corrosion more than outweighs the disadvantage.

2.2.6.3 Laminated Construction

A new departure in transmission line construction, designed to give small attenuation, is suggested by CLOGSTON [264]. The inner conductor of a coaxial line, for example, consists of a central insulating core surrounded by numerous thin cylindrical layers of metal and dielectric disposed alternately. The whole is enclosed in the usual cylindrical outer shield with the usual insulating material separating the inner stack from the shield. The stack has a certain average permittivity for transverse electric fields and it is shown that the wave, and accompanying currents, penetrate most deeply into the stack (this being required to reduce the effective resistance) if the wave velocity on the line is equal to that associated with this average permittivity. This may be brought about by a suitable choice of the main bulk of insulating material.

A transmission line completely filled with laminae is discussed and an analysis is given of the modes of transmission of such a system and of the problem of terminating the line. The theory of planar arrangements is developed for both infinitely thin laminae and for laminae of finite thickness. Experimental work on a partly laminated line is reported by BLACK et al. [265].

2.3 HELICAL LINES

Rectilinear transmission lines, operating in the principal mode, have the outstanding advantage over other types that the frequency characteristic is virtually perfect. The major drawback with such lines is the very limited range of characteristic impedances which are available if the transverse dimensions are not to be impracticably large (see graph Appendix II). Also, when long delay times are required, the corresponding lengths of cable may be inconveniently great in some applications. Transmission cables have accordingly been developed in which the inner conductor is in the form of a closely wound, single layer, spiral; this helix is enclosed within the normal cylindrical sheath. In such an arrangement the series inductance L is greatly increased with a consequent increase in Z_0 and T as defined by equations 2.14 and 2.8.

Propagation no longer occurs in a true TEM mode, but, provided the frequency is not too high, we can apply quasi-stationary theory and find a frequency independent inductance per unit length applicable to propagation in a "pseudo-principal" mode. All the results of § 2.2.5 continue to obtain.

2.3.1 Formula for Inductance per Unit Length

Our first task is to establish a formula for the inductance per unit length of an infinitely long uniform circular helix which is surrounded by a coaxial cylindrical shield; for generality we shall also suppose that a circular conducting rod (or tube) is inserted along the axis within the helix (Fig. 2.14). There has been some discussion (WINKLER [266], FRANKEL [267]) on the matter and an analysis is reproduced here in order that the range of validity of the result may be made clear.

Suppose that the helix is equivalent to a thin current sheet which conducts only in a direction ψ to the x-axis. The surface current density then has an axial component J_{x2} and an azimuthal component $J_{\theta2}$ where the suffix 2 refers to the helix. In the millimicrosecond range the currents in the rod are confined to its surface (equation 2.43) and those in the shield to its inner surface; we can accordingly assign axial surface current densities J_{x1}, J_{x3} and azimuthal current densities $J_{\theta 1}$, $J_{\theta 3}$ to these two conductors respectively. By symmetry all currents are independent of the azimuthal angle and we also suppose that they are independent of the axial distance.

(a) *Axial fields due to azimuthal currents*—In the quasi-stationary case the axial magnetic field H_x inside an infinitely long cylindrical

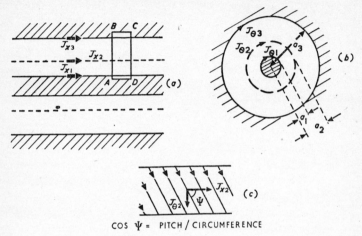

Fig. 2.14. Helix with outer shield and inner rod.
(a) Longitudinal section.
(b) Transverse section.
(c) Components of current density on equivalent slant-conducting sheet.

current sheet, in which the current flows in the azimuthal direction, is uniform and equal to the surface current density (M.K.S. units); the field outside is zero.

If the rod is assumed to be perfectly conducting the electric field inside it is everywhere zero hence

$$0 = \oint E \cdot ds = \mu \iint \frac{\partial H_x}{\partial t} dS \qquad (2.46)$$

where the surface integration is performed over the cross-section of the rod and the line integral is taken round the circumference. Since H_x is uniform, and not constant in time, its value must therefore be zero; thus, if r is radial distance we have

$$H_x = J_{\theta 1} + J_{\theta 2} + J_{\theta 3} = 0 \quad 0 < r < a_1 \qquad (2.47)$$

In the remaining regions the following values of H obtain

$$H_x = J_{\theta 2} + J_{\theta 3} \quad a_1 < r < a_2$$
$$H_x = J_{\theta 3} \quad a_2 < r < a_3$$
$$H_x = 0 \quad a_3 < r$$

since we are now outside all the sheets.

If the shield is also supposed to be perfectly conducting the line integral of the electric field taken round the circumference vanishes; accordingly, for the surface integral (similar to equation 2.46) we have

$$\mu \frac{\partial}{\partial t} \int_0^{a_3} 2\pi r H_x dr = 0 \qquad (2.48)$$

On inserting the values of H_x above appropriate to the several ranges of r we find

$$J_{\theta 2}(a_2^2 - a_1^2) + J_{\theta 3}(a_3^2 - a_1^2) = 0 \qquad (2.49)$$

The current I in which we are ultimately interested is the total axial current along the helix. The following relations obtain (fig. 2.13(c))

$$I = 2\pi a_2 J_{x2} \qquad (2.50)$$

$$\frac{J_{\theta 2}}{J_{x2}} = \tan \psi = 2\pi a_2 n \qquad (2.51)$$

where n is the number of turns per unit length for a single layer winding. From these last two relations we can find $J_{\theta 2}$ in terms of I; $J_{\theta 3}$ then follows from relation 2.49 and $J_{\theta 1}$ from equation 2.47. The values of H_x for all values of r are thus determined in terms of I.

(b) *Azimuthal fields due to axial currents*—In the quasi-stationary case the azimuthal field H_θ at radius r is given by $I'/2\pi r$ where I' is the total axial current enclosed within the circle radius r. Thus

$$H_\theta = 0 \quad 0 < r < a_1$$

since there is no current inside the rod;

$$H_\theta = \frac{a_1 J_{x1}}{r} \quad a_1 < r < a_2$$

$$H_\theta = \frac{a_1 J_{x1} + a_2 J_{x2}}{r} \quad a_2 < r < a_3$$

Since there is no accumulation of charge inside the system we must have at any cross-section

$$2\pi(a_1 J_{x1} + a_2 J_{x2} + a_3 J_{x3}) = 0$$

i.e.

$$H_\theta = 0 \qquad a_3 < r$$

Let us now consider the rectangular path depicted in fig. 2.14(a) which lies entirely in a plane containing the axis. Once again we have

$$\oint E \cdot ds = \mu \int\int \frac{\partial H_\theta}{\partial t} dS \qquad (2.52)$$

In the conductors $E = 0$ thus the paths BC and DA contribute nothing to the line integral. For currents having axial and azimuthal components the vector potential has no radial component and accordingly a potential difference along AB and along CD is unambiguously defined. We can suppose that the rod and shield are both connected to earth at one end of the system so that, provided* the length of the system is much less than the free space wavelength, we can regard the whole of the shield and the whole of the rod as being at the same potential (we are virtually assuming that propagation in the coaxial system formed by the rod and shield is infinitely rapid). The line integrals along AB and CD accordingly also vanish and we have

$$0 = \mu \frac{\partial}{\partial t} \int\int H_\theta ds = \mu \frac{\partial}{\partial t} \int_{a_1}^{a_2} H_\theta dr \qquad (2.53)$$

for unit distance in the axial direction. On inserting the appropriate values of H_θ the following relation emerges

$$a_1 J_{x1} \ln \frac{a_3}{a_1} + a_2 J_{x2} \ln \frac{a_3}{a_2} = 0 \qquad (2.54)$$

From relation 2.50 we find J_{x2} in terms of I and 2.54 then gives J_{x1}. The values of H_θ for all values of r are thus determined in terms of I.

(c) *Determination of inductance*—The self inductance L is related to the total magnetic stored energy by the equation

$$\tfrac{1}{2}LI^2 = \tfrac{1}{2}\mu \int\int\int H^2 d\tau \qquad (2.55)$$

---

* If either the inner rod or the outer shield is absent this proviso may be withdrawn.

In order to obtain the inductance per unit length the volume integral is taken over unit axial distance and becomes

$$\mu\pi \int_0^\infty rH^2 dr$$

Now $H^2 = H_x^2 + H_\theta^2$ and on substituting the values of the field components already found the following expression for L results

$$L = 4\kappa_m \pi^2 n^2 \frac{(a_3^2 - a_2^2)(a_2^2 - a_1^2)}{a_3^2 - a_1^2} \times 10^{-7}$$

$$+ 2\kappa_m \frac{\ln \dfrac{a_3}{a_2} \ln \dfrac{a_2}{a_1}}{\ln \dfrac{a_3}{a_1}} \times 10^{-7} \quad \text{H/m} \qquad (2.56)$$

This expression contains the familiar formula for a long solenoid reduced by the shorted turn effects of the rod and shield; there is also an added contribution given by the "cylindrical" inductance of the helix associated with the shield in parallel with that due to the helix in association with the rod.

If the rod is absent we put $a_1 = 0$ and if there is no shield $a_3 \to \infty$ in the expression 2.56.

For the current sheet approximation to apply in practice, it is necessary that the radius of the wire of which the helix is composed and the distance between adjacent turns be both much less than the spacings $a_3 - a_2$ or $a_2 - a_1$. If a low value of inductance is desired, i.e. if the pitch required turns out to be not very much less than the diameter $2a_2$, then a multiple start winding must be employed the pitch of which is equal to $1/n$.

Further points are listed by WINKLER:

(a) The relation 2.56 shows that for a given shield radius a_3 the inductance is a maximum when $a_2 = 0.707\, a_3$ (the contribution of the cylindrical inductance is neglected and no inner rod is present).

(b) If the shield is in the form of a braid some penetration of field occurs and the effective diameter is somewhat greater than if the shield were solid.

(c) The current tends to flow on the inner surface of the helix; thus, when the wire thickness is appreciable, the internal radius, rather than the mean radius of the helix should be taken for a_2.

2.3.2 Capacitance per Unit Length

If the wire radius and the distance between turns is much less than the spacing between the helix and the shield and that between the helix and the rod, then the capacitance per unit length is given, with good accuracy, by the usual formula for concentric cylinders namely

$$C = \frac{\kappa_e}{18 \times 10^9} \left\{ \frac{1}{\ln \frac{a_3}{a_2}} + \frac{1}{\ln \frac{a_2}{a_1}} \right\} \quad \text{F/m} \qquad (2.57)$$

When the wire is of appreciable thickness the external radius of the helix may be substituted for a_2 in the first term and the internal radius in the second. Again, if the pitch is long and the wire thin, a multiple start winding must be used if this formula is to apply. WINKLER points out that, in the absence of the inner rod and neglecting the cylindrical inductance, the characteristic impedance is a maximum, for a given value of a_3, when $a_2 = 0.486\ a_3$. We then have

$$Z_0 = An a_3 \sqrt{\frac{\kappa_m}{\kappa_e}}$$

where the constant A is equal to 96 if wire size and other factors are neglected; in practice a value $A = 90$ usually obtains.

When the impedance is required to be low the diameter of the coil will turn out to be not much less than that of the shield (or not much greater than that of the rod). FRANKEL has given a tentative analysis of the case when the wire diameter and the distance between turns is comparable with or greater than the separation between the helix and the shield. It may be said that the waves travel over the wire with a velocity approximately equal to the free-space value and hence have a resultant velocity along the axis of 3×10^8 $\cos \psi / \sqrt{\kappa_m \kappa_e}$ m/s. OGLAND [268], among others, suggests that the impedance is given more nearly by the expression for a straight wire parallel to a conducting plane rather than by relations 2.56 and 2.57.

2.3.3 Phase Distortion

Helical lines suffer from phase distortion at high frequencies due to the variation of inductance with frequency and to the self capacitance between adjacent turns. (PALERMO [269]).

2.3.3.1 Variation of Inductance with Frequency

In the quasi-stationary derivation of the self inductance given above the current was assumed to be independent of position along the axis of the helix. When the frequency is high, such that the wavelength along the helix in the axial direction is not very much greater than the diameter, the inductance falls off since it is made up by coupling between turns which are no longer all in phase. L. H. PORITSKY and Mrs. M. H. BLEWETT [270, 271] consider the usual slant conducting current sheet and apply MAXWELL's equations to the regions inside and outside the helix (the influence of the rod and shield is not taken into account*). It is found that

$$\frac{L}{L_0} = 2I_1(\alpha)K_1(\alpha) \qquad (2.58)$$

where L_0 is the low frequency inductance per unit length, I_1 and K_1 are modified Bessel functions of the first and second kind respectively, and $\alpha \equiv 2\pi a/\lambda$ a being the coil radius and λ the wavelength on the helix measured in the axial direction. Some numerical values are tabulated below.

When α is small compared to unity it is very useful for practical purposes to express relation 2.58 in the easily memorized approximate form (LEWIS [273])

$$\frac{L}{L_0} \simeq 1 - 25 \left(\frac{a}{\lambda}\right)^2 \qquad (2.59)$$

The extent of the agreement between relations 2.58 and 2.59 is indicated in Table 2.4; the correspondence is close for $\alpha < 0.5$.

The effect of amplitude distortion and phase distortion on the output pulse shape in any system is discussed by KALLMANN [274]. In the case of a helical delay line, phase distortion, rather than amplitude distortion due to frequency dependent attenuation, may well be the limiting factor and a convenient criterion is that the total phase

---

* A wave analysis of a spiral line within a metal tube is given by LOSHAKOV and OLDEROGGE [272].

change along the whole line suffered by the highest relevant frequency component should not differ by more than $\frac{1}{2}$ radian from the phase change which would occur were the phase characteristic perfectly linear. This criterion may be expressed more explicitly as follows.

TABLE 2.4. Variation of self inductance with frequency

α	$2I_1(\alpha)K_1(\alpha)$	$1 - 25(a/\lambda)^2$
0·00	1·000	1·000
0·05	0·996	0·998
0·10	0·986	0·994
0·25	0·944	0·960
0·50	0·854	0·842
0·75	0·764	0·644
1·00	0·680	
1·50	0·545	
2·00	0·445	

Let T be the time delay per unit length at a (high) frequency ω which differs from the value T_0 at medium frequencies. The phase error at the angular frequency ω after traversing a length l of line must not exceed $\frac{1}{2}$ radian and the useful upper limit, neglecting resistive losses, may therefore be defined by

$$\omega_{\max} l (T - T_0) = \pm \tfrac{1}{2} \tag{2.60}$$

Now $T = \sqrt{LC}$ and if we suppose that C is independent of frequency while L depends on frequency in some manner this relation becomes

$$\omega_{\max} l T_0 \left(\sqrt{\frac{L}{L_0}} - 1 \right) = \pm \tfrac{1}{2} \tag{2.61}$$

If we now write $L = L_0 + \delta L$ on the supposition that the fractional variation in inductance is small we have

$$\omega_{\max} l T_0 \frac{\delta L}{L_0} = \pm 1 \tag{2.62}$$

If the variation in inductance is due to the effect discussed above, substitution for L/L_0 from equation 2.59 into 2.61 yields

$$\omega_{\max} T_0 (la^2)^{\frac{1}{3}} \simeq 3 \tag{2.63}$$

provided $\omega_{max}T_0 a < 0.5$; (the relation $1/T_0 = \omega/2\pi\lambda$ has been invoked). The result may be written in the form

$$\omega_{max} \simeq \frac{3}{T_0 l}\left(\frac{l}{a}\right)^{\frac{2}{3}} \qquad (2.64)$$

which relates the upper useable frequency limit to the total delay time $T_0 l$ and to the ratio of the length of the helix to its radius.

2.3.3.2 Effect of Self Capacitance

The effect of the self capacitance between turns has been analysed by SHAW [275] and others using a lumped equivalent network. If C' is the self capacitance between two adjacent turns one would expect that the series impedance per turn would consist of the inductance of one turn, namely L/n, in parallel with the lumped capacitance C'. At frequencies below the self resonant frequency

$$\omega_0 = \sqrt{\frac{n}{LC'}} \qquad (2.65)$$

the combination still appears inductive and it follows at once that

$$\frac{L}{L_0} = \frac{1}{1 - (\omega/\omega_0)^2} \qquad (2.66)$$

This result has been shown by LEWIS to be correct, provided $\omega \ll \omega_0$, when the truly distributed nature of the self capacitance is taken into account.

It is noticed, on comparing the results 2.59 and 2.66, that the two effects act in opposition to each other. Mutual compensation may therefore be aimed at with a view to extending the frequency range over which the effective inductance, and therefore the delay time and characteristic impedance, are independent of frequency. Work has been done to this end. KALLMANN [274] and ERICKSON and SOMMER [276] give theoretical and practical accounts of the use of insulated metal strips which are placed close to the coil in order to increase the self capacitance of the winding.

WEEKES [277] has employed lumped correcting networks at the

terminations of the line in order to reduce phase distortion and DI TORO [278] has used a bank wound multi-layer solenoid.

2.3.4 General.

It may be pointed out that the shorted turn effect of the rod and/or shield on the inductance may be reduced by slitting the rod and shield longitudinally, in several places round the circumference, throughout the entire length.

When high values of impedance and delay time are required two helices may be employed, wound in opposite directions one on top of the other. The two coils form the go and return circuits and, if the coupling is perfectly tight, the inductance per unit length is four times that for one solenoid alone; the shunt capacitance is simply that between the two windings. The system is no longer completely unbalanced with respect to earth and no earth connection is common to the two extremities of the line. Its use is therefore normally limited to pulse shaping applications in which the input and "output" are at the same end.

Exact matching at the end of a helical line presents a problem. The inductance per unit length falls off towards the ends to one half the value applicable to the central region of a long solenoid. If reflection is to be avoided the characteristic impedance must, of course, remain constant throughout the whole length; considerable improvement is obtained if the winding is flared near the ends so as to raise L back to the normal value or by increasing the spacing between the helix and the shield (or rod) in such a manner that the capacitance decreases according to the same law thus keeping the ratio L/C constant. Even when such steps are taken reflection will still occur at the higher frequencies owing to the improper matching of the fields. When it is desired to pass from a coaxial line into a coaxial cable the transition should be made smoothly; a transition section may be constructed by cutting a slot of gradually increasing pitch in a circular hollow tube which forms the inner conductor of the system. Such transitions, developed for use in the microwave region but adaptable for our purposes, are discussed by LUND [279]; a change in characteristic impedance will usually be involved and the electrical length of the transition must therefore be several times the pulse length (see Chapter 3).

Some data on commercially available cables are given in the table below. A helical conductor is embedded in polythene dielectric which fills the interior of a braided outer shield; the whole is mechanically flexible but sharp bends should be avoided.

TABLE 2.5. Data for Delay Cables

T.C.M. Type		Z_5	Z_3	Z_1
Characteristic Impedance, ohms		130	330	550
Length, feet, for 100 mμsec. delay		14·3	5·8	3·6
Capacitance, pF per foot		55	50	50
Overall diameter, inches		0·35	0·61	1·0
Attenuation db for 100 mμsec. delay	1 Mc/s	0·33	0·10	0·05
	10 Mc/s	1·15	0·35	0·18
	100 Mc/s	3·6	1·28	0·83

These cables do not appear to be compensated for delay distortion and the attenuation figures seem to give an optimistic impression as regards the useful high-frequency limit for long delays.

Further information on helical lines may be found in papers by ESSEN [280], HODELIN [281], KALLMANN [282], RUBEL et al. [283] and ZIMMERMAN [284].

2.4 LUMPED DELAY LINES

Our chapter on transmission systems may close with a brief account of the low-pass filter or delay line composed of lumped capacitors and inductors. The performance of such a line is inferior to that of the helical line and a lumped network should never be employed when a continuous structure could be used instead. In circuit design, however, we are continually being faced with the existence of small lumped stray capacitances to earth, notably the grid and anode capacitances of valves, and filter theory shows how a considerable measure of compensation can be achieved by the addition of series inductances. Much information is available in the literature; the Bibliography may be consulted and also papers by HEBB et al. [285], KALLMANN et al. [286], THOMSON [287], TREVOR [288], and WHEELER and MURNAGHAN [289].

2.4.1 Constant-k Filters

The network of fig. 2.15(a) can propagate signals, in either direction, with a phase change β per section at angular frequency* ω given by

$$\cos \beta = 1 - 2 \left(\frac{\omega}{\omega_c}\right)^2 \tag{2.67}$$

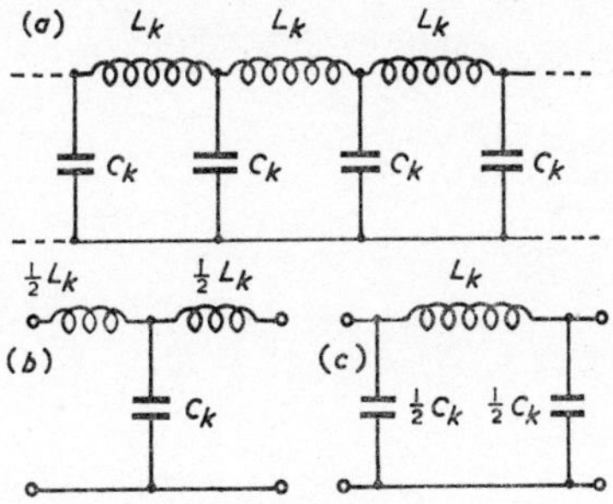

Fig. 2.15. (a) Constant-k low-pass filter, (b) T-section, (c) Π-section.

The quantity β is real when the frequency is less than the cut-off frequency

$$f_c = \frac{\omega_c}{2\pi} = \frac{1}{\pi\sqrt{L_k C_k}}$$

If $\omega/\omega_c \ll 1$ the expression for the time delay T per section is†

$$T = \sqrt{L_k C_k} \left[1 + \frac{1}{6}\left(\frac{\omega}{\omega_c}\right)^2 - \ldots\right] \tag{2.68}$$

At frequencies above the cut-off no propagation occurs; signals are strongly reflected at the input end and are attenuated rapidly down the line. When a very few sections are employed the cut-off frequency fixes the upper limit of usefulness but when the number of sections is large phase distortion will be the limiting factor; such distortion will exceed a tolerable amount (see § 2.3.3.1) at a frequency considerably below the cut-off dependent on the number of sections

---

* For a Laplace transform treatment see DAWES *et al.* [290].
† We have here taken the time delay T as being equal to β/ω. This is correct when a sinusoidal signal of one frequency only is concerned but when we are dealing with a band of frequencies the group time delay $d\beta/d\omega$ may be more appropriate. In the latter case the number 6 in relation 2.68 is replaced by the number 2. The distinction is of little practical importance however.

used (see fig. 2.17(b) for the case $m = 1$). We shall usually suppose, however, that the cut-off frequency may be taken as the limit in practical cases but the above reservation must always be borne in mind.

The network shown in fig. 2.15(a) may be regarded as being made up of a number of symmetrical "T" sections, (fig. 2.15(b)) or of "Π" sections (fig. 2.15(c)) connected in cascade. In the body of the line the two are indistinguishable, as far as propagation is concerned, but a distinction arises when we come to find the impedance by virtue of the differing form of the terminating sections. The T-filter finishes at each end with a series inductance $L_k/2$ and the Π-section with a shunt capacitance $C_k/2$. In our applications the stray capacitances of the source and load will automatically provide a shunt capacitance at each termination and accordingly the Π-type of section will be encountered more often in practice.

The filter possesses a certain characteristic impedance (often termed the iterative impedance) which is purely resistive, in the pass band, but dependent on frequency. Tor T-sections we have

$$Z_0 = \sqrt{\frac{L_k}{C_k}\left[1 - \left(\frac{\omega}{\omega_c}\right)^2\right]} \tag{2.69}$$

$$\simeq \sqrt{\frac{L_k}{C_k}}\left[1 - \frac{1}{2}\left(\frac{\omega}{\omega_c}\right)^2 + \ldots\right] \quad \text{provided } \omega/\omega_c \ll 1$$

and in the case of a Π-section

$$Z_0 = \sqrt{\frac{L_k}{C_k\left[1 - \left(\frac{\omega}{\omega_c}\right)^2\right]}} \tag{2.70}$$

$$\simeq \sqrt{\frac{L_k}{C_k}}\left[1 + \frac{1}{2}\left(\frac{\omega}{\omega_c}\right)^2 \ldots\right] \quad \text{when } \omega/\omega_c \ll 1$$

If one or more sections are terminated by a resistance equal to the characteristic impedance no reflection occurs and the input impedance is equal to the characteristic impedance, as in the case of a transmission line. A resistance which varies with frequency in the prescribed manner is difficult to devise and, for the case of a single section only, GIACOLETTO [291] has worked out certain optimum values for the termination when the latter consists of a frequency independent resistance R at both ends. Thus, in the case of a T-section, for the input impedance to be as uniformly resistive as possible we should make $R = 0.75\,Z_0$; for minimum reactive impedance $R = 0.95\,Z_0$, and for the phase shift to be as linear as possible $R = 0.97\,Z_0$. For Π-sections these figures read 1·5, 2·06 and 1·65 respectively. The optimum value is defined as that which gives a minimum value, over the whole pass band, to the squares of the deviations from the desired condition.

2.4.2 Derived Filters.

Phase distortion in the simple constant-k type filter may be reduced by employing the so called m-derived form shown in fig. 2.16(a). The

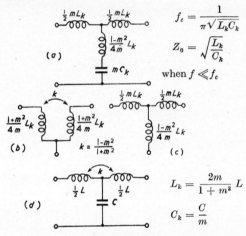

Fig. 2.16. Series m-derived T filter section.

extra inductance in the shunt arm is related to the other reactances by the parameter m in the manner indicated. The cut-off frequency is again equal to $1/\pi\sqrt{L_k C_k}$ and the characteristic impedance is also given by relation 2.69 for all values of ω and m. The expression for the phase change β per section now becomes

$$\cos \beta = \frac{1 - (1 + m^2)\left(\dfrac{\omega}{\omega_c}\right)^2}{1 - (1 - m^2)\left(\dfrac{\omega}{\omega_c}\right)^2} \qquad (2.71)$$

On putting $m = 1$ the circuit of fig. 2.16(a) reduces to that shown in fig. 2.15(b) and the expression 2.71 for β reduces to relation 2.67. The variation, with frequency, of group delay and of phase error for

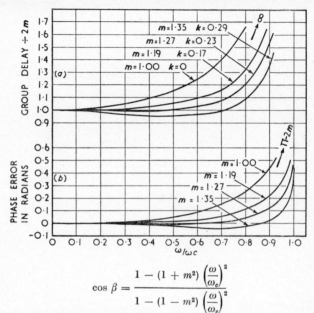

$$\cos \beta = \frac{1 - (1 + m^2)\left(\dfrac{\omega}{\omega_c}\right)^2}{1 - (1 - m^2)\left(\dfrac{\omega}{\omega_c}\right)^2}$$

(a) Group delay, per section

$$\frac{d\beta}{d\left(\dfrac{\omega}{\omega_c}\right)} = \frac{2m}{\sqrt{1 - \left(\dfrac{\omega}{\omega_c}\right)^2}\left[1 - (1 - m^2)\left(\dfrac{\omega}{\omega_c}\right)^2\right]}$$

(b) Phase error, per section

$$= \beta - \left.\frac{d\beta}{d\left(\dfrac{\omega}{\omega_c}\right)}\right|_{\omega = 0} \cdot \left(\frac{\omega}{\omega_c}\right) = \beta - 2m\left(\frac{\omega}{\omega_c}\right)$$

For constant-k filter $m = 1$

Fig. 2.17. Filter characteristics.

various values of m, is plotted in fig. 2.17; the value $m = 1.27$ is a good compromise (for both sets of curves).

A value of m greater than unity corresponds to a negative inductance in series with the capacitance. This may be realized physically by allowing inductive coupling to take place between the two half coils (across the "T") as follows from the equivalence of figs. 2.16(b) and (c). For $m > 1$ the coefficient of coupling k* must be negative, i.e. coils lying on the same former should be wound in the same direction.

In the case $m = 1\cdot27$, which corresponds to $k = 0\cdot23$, the fundamental line parameters may alternatively be expressed in terms of the actual L and C (fig. 2.16(d)) as follows

$$\left. \begin{array}{ll} \text{Characteristic impedance} & Z_0 = 1\cdot11 \sqrt{\dfrac{L}{C}} \\[6pt] \text{Delay per section} & T = 1.11 \sqrt{LC} \\[6pt] \text{Cut-off frequency} & f_c = \dfrac{0.36}{\sqrt{LC}} \end{array} \right\} \qquad (2.72)$$

The simplest method of terminating the network is indicated in fig. 2.18(a). A better match is obtained by inserting an m-derived half section, with $m = 0\cdot6$, as shown in fig. 2.18(b) (the value

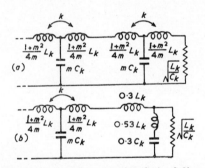

Fig. 2.18. Termination of m-derived filter.
(a) Simple resistance termination.
(b) Termination via $m = 0\cdot6$ derived half-section.

$m = 1\cdot27$ obtaining in the rest of the line). The coils in the terminating half-section are not coupled to one another and the capacitance can be made variable so that an optimum setting may be arrived at.

A line composed of bridged T-sections (fig. 2.19(a)) may be employed to give reduced phase distortion. FERGUSON [292] shows how higher order frequency terms in the expression for the time delay may be eliminated by adding capacitances according to

---

* The quantity k denoting the coefficient of coupling has nothing to do with the suffix k referring to constant-k sections.

fig. 2.19(b), and Golay [293] has obtained good results by connecting capacitances between alternate tie points as indicated in fig. 2.19(c). The bridging capacitors have little influence on the fundamental line

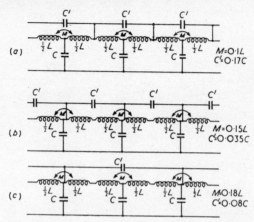

Fig. 2.19. m-derived filters with bridging capacitors.

parameters of the three arrangements depicted in fig. 2.19; the following relations again apply, at least approximately,

$$\left. \begin{array}{ll} \text{Characteristic impedance} & Z_0 = \sqrt{\dfrac{L+2M}{C}} \\ \text{Delay per section} & T = \sqrt{(L+2M)C} \\ \text{Cut-off frequency} & f_c = \dfrac{1}{\pi\sqrt{(L-2M)C}} \end{array} \right\} \quad (2.73)$$

3
TRANSFORMERS

3.1 INTRODUCTION

THE principle of impedance matching (FANO [301]) is widely applied when it is desired to transmit maximum power from a generator into a load. In the simplest case the internal impedance of the generator, Z_1 say, and the impedance of the load, Z_2 say, are both pure resistances. It is easily shown that, for a given generator e.m.f., the power dissipated in the load is a maximum when $Z_1 = Z_2$. The two impedances are different in general and a transformer is accordingly inserted. The secondary to primary turns ratio r is chosen such that, with the load connected across the secondary, the input impedance on the primary side has the value Z_1; for an ideal transformer $r = \sqrt{Z_2/Z_1}$.

When the system includes a length of transmission line, of characteristic impedance Z_0, we have already seen that reflection occurs at the receiving end unless the line is terminated by a resistance Z_0. The condition of no reflection is thus, as expected, identical with that for complete power transfer. In general it will be necessary to employ two transformers, one at the sending end to match the internal impedance of the generator to the characteristic impedance of the line, and another at the receiving end to match the load to the line.

In the sort of equipments we have in mind the securing of optimum power transfer is usually a secondary consideration. It is much more important that subsidiary pulses, produced by multiple reflection of the main pulse at the terminations, should be avoided. Ideally, of course, it should only be necessary to match the line at one end in order to eliminate multiple reflections, but, since the matching can never be perfect and is often comparatively poor in practice, it is generally safer to terminate the line at both ends in the nominally correct manner.

Again we usually wish to obtain maximum voltage amplitude across the load and when the impedance is high (the grid-cathode circuit of a valve for example) one is tempted to omit the terminating resistor altogether (MARSHALL [302]) and obtain a factor of two improvement in signal amplitude over the matched case as shown in § 2.2.5. This procedure is open to the objections that (i) multiple reflections will occur unless the line is accurately matched at the generator end, and (ii) the effect of the inevitable capacitance of the load will be much more pronounced than would be the case if the correct resistive termination were added (see waveforms given in fig. 2.13).

3.2 ELEMENTARY MATCHING NETWORKS

Before considering pulse transformers proper it is worth pointing out that certain simple resistance networks provide an easy means of impedance matching when the avoidance of reflections is important even at the expense of a loss in signal amplitude. We suppose that it is required to match two transmission lines of characteristic impedances Z_{01} and Z_{02} where $Z_{01} < Z_{02}$. The arrangements shown in fig. 3.1 give a match on looking inwards in the direction indicated.

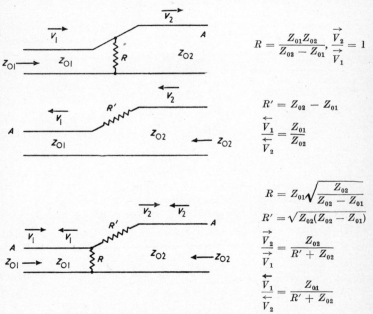

Fig. 3.1. Simple matching by resistors. In all cases $Z_{01} < Z_{02}$.

The lines are assumed to be properly terminated at the ends A. The use of the smallest number of resistances gives least voltage attenuation in all cases.

3.3 LUMPED PULSE TRANSFORMERS

The properties of lumped pulse transformers have been thoroughly investigated in the microsecond range. The reader may consult papers by HADLOCK [303], MAURICE and MINNS [304], MELVILLE [305], MOODY [306, 307], and the book by GLASCOE and LEBACQZ (see Bibliography). It has been found by MOODY *et al.* [308] that existing design theory is applicable in the millimicrosecond region.

3.3.1 Equivalent Circuit.

A typical approximate equivalent circuit is shown in fig. 3.2(a). A common earth connection between the input and the output sides

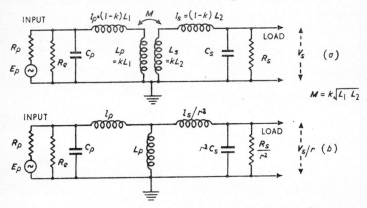

Fig. 3.2. (a) Approximate equivalent circuit of pulse transformer.
(b) Transformed equivalent of the above.

is assumed for convenience and the capacitance between the windings is neglected (if an electrostatic screen is interposed between primary and secondary this assumption will be quite justifiable). L_1 is the total primary inductance which may be imagined to consist of the sum of an effective primary inductance $L_p = k L_1$ plus a leakage inductance $l_p = (1 - k)L_1$ where k is the coefficient of coupling. The secondary likewise has a total inductance L_2 made up of an effective part $L_s = kL_2$ plus a leakage inductance $l_s = (1 - k)L_2$. C_p and C_s represent the stray capacitances to earth of the primary and secondary windings respectively and R_e is a shunt resistance to which the losses in the core material are equivalent (the resistance losses in the coils themselves are usually negligible). R_p and R_s are the resistances of the source and of the load respectively.

The effective inductances L_p and L_s by definition form an ideal transformer with perfect coupling, the mutual inductance M being given by $M = \sqrt{L_p L_s} = k\sqrt{L_1 L_2}$. If r is the secondary to primary turns ratio of this idealized part of the transformer it is convenient to transfer all impedances to the primary side; performance may then be predicted by considering the equivalent circuit of fig. 3.2(b) so derived.

3.3.2 Limitations on Performance.

The high frequency response of the circuit is limited by the leakage inductances and shunt capacitances (see RUDENBERG [309, 310, 311]): the physical size must therefore be small and high permeability material used for the core in order that k may be made as near unity as possible. The high frequency performance is improved if the circuit values are arranged such that the leakage inductances and the shunt capacitances form a low-pass filter section (see § 2.4.1); the shunt inductance L_p is high and plays no part at the higher frequencies. The transformer should be designed such that the resistances R_p and R_s/r^2 are each equal to the characteristic impedance of the filter section. Smooth windings are necessary in order to avoid internal reflections and the self capacitance of the windings should be as small as possible.

Magnetic tape, 0·001 in. thick, may be used for the core and MOODY and his collaborators have found that the effective or pulse permeability is given by

$$\kappa_m(\tau) = 6{\cdot}2\,\frac{\sqrt{\rho\tau\kappa_m(\infty)}}{d} \tag{3.1}$$

when the flux penetrates only partially into the material. Here $\kappa_m(\infty)$ is the permeability for complete penetration, ρ is the core resistivity in microhms cm, d is the thickness of the lamination in thousandths of an inch and τ the pulse duration in microseconds. For a typical tape $\kappa_m(\infty) = 800$, $\rho = 50$, $d = 1$ and accordingly $\kappa_m(\tau) = 240$ at $\tau = 10$ mμsec. Various ferroxcube cores function satisfactorily but have been found to be inferior to thin mumetal whorls.

A step function voltage input E_p will be differentiated, to a first approximation, with a time-constant equal to

$$\frac{L_p \left(R_p + \dfrac{R_s}{r^2} \right)}{R_p R_s / r^2} \quad (3.2)$$

where the resistance R_e is neglected. The primary inductance must therefore be made high enough to suit the length of the pulse which the transformer is required to pass. A typical 75 Ω to 200 Ω step-up auto transformer transmits a 300 mμsec. rectangular pulse with a decay in amplitude of only 2·5% and with a rise time of 1 mμsec.

It must be pointed out that special magnetic alloys must be handled with care, avoiding heat or mechanical working, if their magnetic properties are not to suffer.

Advantages possessed by lumped pulse transformers over the tapered lines about to be described include (i) extremely small physical size, (ii) the facility of optional phase inversion, and (iii) the isolation of the secondary from the primary circuit as far as the mean potentials of the two are concerned.

3.4 TAPERED LINES

We have seen in § 2.2.5 how the time delay property possessed by a transmission line, together with reflection phenomena, are applied. The fact that a line possesses a characteristic impedance, which is purely resistive although the system is completely lossless (in the ideal case), suggests that a line may form the basis of an impedance transformer.

3.4.1 Quarter-wave Transformer.

This principle has been widely used in the field of short wave radio transmission. When it is desired to match a line of characteristic impedance Z_{01} into a resistive load (or a correctly terminated length of a different line) of impedance Z_{02}, a section of line of characteristic impedance $\sqrt{Z_{01} Z_{02}}$ and one quarter of a wavelength long may be inserted. It can be shown that there is no resultant wave reflected back towards the sending ending due to mutual cancellation of the reflections at the two impedance discontinuities.* The device thus

---

* Throughout the following sections we shall omit specific reference to the discontinuities discussed in § 2.2.4.2; the possible effect of these must always be borne in mind however.

provides a match which is complete at the particular frequencies for which the section is an odd number of quarter wavelengths long. It is known that the frequency band over which the resultant reflection is sensibly zero increases as more quarter-wave sections are added provided the increments in $\ln Z_0$, as we pass through the various sections from input to output, are arranged to be proportional to the binomial coefficients

$$1,1 \quad 1,2,1 \quad 1,3,3,1 \quad 1,4,6,4,1 \quad \text{etc}$$

In the limit of large numbers these coefficients give rise to a Gaussian error function distribution and we should thus expect good results to be obtained when $\ln Z_0$ changes smoothly according to a Gaussian law (see § 3.4.3).

3.4.2 Analysis of Smooth Tapered Transmission Lines.

Many workers have directed their attention to the case of a transmission line in which the inductance and capacitance per unit length vary with position along the line. The analyses are based on circuit theory and for this to be valid the electromagnetic fields must be essentially of the principal mode type, i.e. the fractional changes in the transverse dimensions, in a distance along the line of the order of the transverse dimensions, must be small. The same applies to helical lines but in addition the wavelength along the helix must be much greater than the diameter, as we have seen, in order that the inductance per unit length may be independent of frequency.

It may be noted at the outset that any tapered line, necessarily of finite length, must be partially reflecting at low frequencies. This is because at sufficiently low frequencies the device is much less than one wavelength long and even the tapered line presents an abrupt change in impedance between the values at the two ends.

We shall restrict ourselves to the case of lossless lines since, as pointed out in § 2.2.6.1, this does not entail an immediate loss of generality. Returning to equations 2.2 and 2.3 we must remember that L and C are now functions of x when performing the elimination of $\bar{I}$ or $\bar{V}$. It is easily shown that

$$\frac{d^2 \bar{V}}{dx^2} - \frac{1}{L}\frac{dL}{dx}\frac{d\bar{V}}{dx} - p^2 LC \bar{V} = 0 \qquad (3.3)$$

$$\frac{d^2 \bar{I}}{dx^2} - \frac{1}{C}\frac{dC}{dx}\frac{d\bar{I}}{dx} - p^2 LC \bar{I} = 0 \qquad (3.4)$$

We may define the nominal characteristic impedance as

$$Z_{0x} = \sqrt{\frac{L_x}{C_x}}$$

This expression is identical with that met with in the case of the uniform line but is now a function of position having the value $Z_{00} = \sqrt{L_0/C_0}$ say at $x = 0$. Similarly the nominal time delay per unit length is defined* by the relation

$$T_x = \sqrt{L_x C_x}$$

3.4.2.1 First Approximation to a Solution

A first approximation to a solution of equations 3.3 and 3.4 has been given by SLATER [312]. Looking back at the solution 2.6 for the uniform line it would be expected that since the parameters vary with x the quantity $\vec{V}_0$ would no longer be constant but should include a function of x. Also, since the delay per unit length is variable in general, it would seem reasonable to replace the exponent $-pTx$ by the expression

$$-p \int_0^x T(x')dx' \tag{3.5}$$

Following these suggestions a trial solution

$$\vec{V}_x = \vec{A}(x) e^{-p\int_0^x T(x')dx'} \tag{3.6}$$

is substituted into equation 3.3. If certain terms in the resulting differential equation are neglected (see below) it is easy to solve for $\vec{A}(x)$. The solution then reads

$$\vec{V}_x = \vec{V}_0 \sqrt{\frac{Z_{0x}}{Z_{00}}} e^{-p\int_0^x T(x')dx'} \tag{3.7}$$

where, as in §§ 2.2.2 et seq., $\vec{V}_0$ is the Laplace transform of the voltage signal at $x = 0$ associated with a wave travelling to the right. There

---

* T is constant for rectilinear tapered lines being equal to the reciprocal of the velocity of electromagnetic waves in the medium. In a helical line, however, both L and C may vary arbitrarly.

also exists an independent solution, similar to 3.7 and to be added to it, corresponding to a wave propagated in the reverse direction i.e. along the negative x-axis. The analogous solutions for the current are obtained simply by interchanging L and C and introducing two new arbitrary constants $\vec{I}_0$ and $\overleftarrow{I}_0$.

In deriving these solutions it has been assumed that

$$\left|\frac{1}{pTL}\frac{dL}{dx}\right| \ll 1, \left|\frac{1}{pTC}\frac{dC}{dx}\right| \ll 1 \qquad (3.8)$$

and

$$\left|\frac{1}{pT}\frac{d^2\sqrt{Z_0}}{dx^2}\right| \ll \left|\frac{d\sqrt{Z_0}}{dx}\right|, \left|\frac{1}{pT}\frac{d^2\sqrt{1/Z_0}}{dx^2}\right| \ll \left|\frac{d\sqrt{1/Z_0}}{dx}\right| \qquad (3.9)$$

To interpret these ineqalities we must digress from the Laplace transform method and consider the sinusoidal case. On putting $p = j\omega$ and $\omega = 2\pi/\lambda T$ where λ is the wavelength along the line the conditions 3.8 become

$$\left|\frac{\lambda}{L}\frac{dL}{dx}\right| \ll 2\pi, \left|\frac{\lambda}{C}\frac{dC}{dx}\right| \ll 2\pi \qquad (3.10)$$

The solution 3.7 ignores reflections and implies that no distortion of pulse shape will occur. The conditions 3.10 show that for this to be the case the fractional change per wavelength in both L and C must be very much less than six for the lowest relevant frequency component in the signal pulse. If both L and C vary thus slowly Z_0 will also vary slowly and the condition 3.9 will usually be automatically satisfied, except possibly just at the ends of the line where dZ_0/dx may be discontinuous; such discontinuities will be taken into account when we come to consider reflections at the terminations.

A relation between the arbitrary voltage and current amplitude constants at $x = 0$ is provided by relation 2.2, and it is found that

$$\frac{\vec{V}_0}{\vec{I}_0} = -\frac{\overleftarrow{V}_0}{\overleftarrow{I}_0} = Z_{00} \qquad (3.11)$$

and

$$\frac{\vec{V}_x}{\vec{I}_x} = -\frac{\overleftarrow{V}_x}{\overleftarrow{I}_x} = Z_{0x} \qquad (3.12)$$

TRANSFORMERS

The ratio of the voltage to the current at any point (for each component wave) is thus equal to the nominal characteristic impedance at the point in question to the degree of accuracy indicated by conditions 3.10. Equation 3.7 shows that, apart from the time delay factor, the voltage amplitude along the line is proportional to the square root of the characteristic impedance, as would be expected from energy considerations in the absence of any reflected waves. The tapered line thus does in fact provide the desired voltage and impedance transformation.

In the range of millimicrosecond pulses the physical length of tapered line transformers is no longer prohibitive, particularly if a helical transmission line is employed. Extremely fast rise times are attainable, with rectilinear lines, since the high frequency performance is the same as that of a uniform line of similar dimensions. Such devices do not require a ferromagnetic core and are perfectly linear; high powers can be readily handled, construction is easy and the tolerances are comparatively wide. The arrangements, however, lack the advantages of optional phase inversion and circuit isolation between input and output, which lumped transformers possess. (A transmission line type of pulse inverter is described in § 3.5).

3.4.2.2. Second Approximation

Having seen what the approximate behaviour of a tapered line will be, let us now seek a more accurate solution to equations 3.3 and 3.4. The total time delay $u(x)$ from the origin to the point x is given by

$$u(x) = \int_0^x T(x')dx' \tag{3.13}$$

i.e.

$$du = T(x)dx$$

If the quantity u, rather than x itself, is taken as the independent variable the basic equations 3.3 and 3.4 become

$$\frac{1}{p^2}\frac{d^2\bar{V}}{du^2} - \frac{1}{p^2}\frac{d\ln Z_0}{du}\frac{d\bar{V}}{du} - \bar{V} = 0 \tag{3.14}$$

$$\frac{1}{p^2}\frac{d^2\bar{I}}{du^2} + \frac{1}{p^2}\frac{d\ln Z_0}{du}\frac{d\bar{I}}{du} - \bar{I} = 0 \tag{3.15}$$

These equations form the starting point of several methods of attack. They cannot, in general, be solved exactly except in the case of the exponential line which is treated in § 3.4.4. Approximate solutions, more accurate than 3.7, can, however, be obtained provided the line parameters are assumed to vary slowly.

FRANK [313] has carried SLATER's result 3.7 to the next order of approximation. A trial solution

$$\bar{V}_u = A \sqrt{\frac{Z_{0u}}{Z_{00}}} e^{\pm pu}[1 + f(u)] \qquad (3.16)$$

where $f(u)$ is a correction to be determined and A is a constant (having different values for the two directions of propagation), is substituted in equation 3.14. It follows that f must satisfy the differential equation

$$\frac{d^2 f}{du^2} \pm 2p \frac{df}{du} + \left[\frac{1}{Z_0} \frac{d^2 Z_0}{du^2} - \frac{3}{2} \left(\frac{1}{Z_0} \frac{dZ_0}{du} \right)^2 \right] \frac{(1+f)}{2} = 0 \quad (3.17)$$

If we now suppose that the correction f is small and slowly varying this reduces to

$$\pm \frac{df}{du} + \left[\frac{1}{Z_0} \frac{d^2 Z_0}{du^2} - \frac{3}{2} \left(\frac{1}{Z_0} \frac{dZ_0}{du} \right)^2 \right] \cdot \frac{1}{4p} = 0 \qquad (3.18)$$

provided

$$|f| \ll 1 \quad \text{and} \quad \left| \frac{d^2 f}{du^2} \right| \ll \left| 2p \frac{df}{du} \right| \qquad (3.19)$$

On integrating we find

$$f = \pm \frac{1}{4p} \left[-\frac{d \ln Z_0}{du} + \frac{1}{2} \int_0^u \left(\frac{d \ln Z_0}{du'} \right)^2 du' + K \right] \quad (3.20)$$

where K is an arbitrary constant. The original variable x may now be reintroduced, and, neglecting second order small quantities, equation 3.16 yields for the complete solution

$$\bar{V}_u = \overrightarrow{V}_u + \overleftarrow{V}_u \qquad (3.21)$$

where

$$\overrightarrow{V}_u = \overrightarrow{V}_0 \sqrt{\frac{Z_{0u}}{Z_{00}}} e^{-pu} \left\{ 1 + \frac{1}{4pT} \frac{d \ln Z_0}{dx} - \frac{1}{4pT_0} \frac{d \ln Z_0}{dx} \bigg|_{x=0} \right.$$
$$\left. - \frac{1}{8p} \int_0^x \frac{1}{T} \left(\frac{d \ln Z_0}{dx'} \right)^2 dx' \right\}$$

$$(3.22)$$

and

$$\overleftarrow{V}_u = \overleftarrow{V}_0 \sqrt{\frac{Z_{0u}}{Z_{00}}} e^{+pu} \left\{ 1 - \frac{1}{4pT} \frac{d \ln Z_0}{dx} + \frac{1}{4pT_0} \frac{d \ln Z_0}{dx} \bigg|_{x=0} \right. \\ \left. + \frac{1}{8p} \int_0^x \frac{1}{T} \left(\frac{d \ln Z_0}{dx'} \right)^2 dx' \right\}$$

(3.23)

The constants A have been expressed in terms of the voltage amplitudes $\overrightarrow{V}_0$ and $\overleftarrow{V}_0$ of the two wave components at $u = x = 0$ (the constants K disappear). T_0 and Z_{00} are the values of the delay per unit length and the nominal characteristic impedance respectively at the origin.

Similar results for the currents, involving constant $\overrightarrow{I}_0$ and $\overleftarrow{I}_0$ are obtained from equation 3.15. The solutions may be written down from 3.22 and 3.23 simply by replacing Z_0 by $1/Z_0$ and Z_{00} by $1/Z_{00}$.

For the solutions to be valid both the inequalities 3.19 must be satisfied. The first condition is readily checked in an actual example since it only implies that the correcting terms in the solutions 3.22 and 3.23 must turn out to be small. The expression of the second condition, in terms of the variable x, is too lengthy to be reproduced here and in any case will be satisfied in practice if the line parameters vary smoothly. Questions of the continuity of T are involved and it might be thought that both L and C should vary smoothly (as required in the analysis summarized in § 3.4.2.1 by the inequalities 3.8). On inspection of equations 3.14 and 3.15, however, it is seen that the quantities L and C appear as a dependent variable only in the ratio $Z_0 = \sqrt{L/C}$ and we would therefore expect that L and C may vary separately in quite an arbitrary manner provided the impedance varies slowly. It can be shown that T need not in fact be a continuous function of x.

The variation of pulse amplitude with time, as observed at a particular point x, u, is obtained from equation 3.21 by taking the inverse Laplace transformation. In addition to the factor representing voltage transformation (in the electrical not the mathematical sense) and the exponential term which embodies the bulk time delay u, we have to consider the correcting terms which involve $1/p$. The effect of these is to distort the pulse by subtracting from it a voltage proportional to the time integral of the input pulse $V_0(t)$.

If the input pulse (once launched in the line) is rectangular in shape the top will be seen to fall* linearly, after the manner shown in fig. 3.4, as the signal passes a fixed observation point on an infinitely long (or perfectly matched) line. A very simple formula applies in the case of the exponential line (see § 3.4.4.1 equation 3.72).

The validity of the result, in any particular example, may be more readily ascertained by invoking the above interpretation than by applying conditions 3.10 which involve a consideration of the low frequency components in the pulse. The rate of taper of the line and the pulse length employed must be such that the predicted distortion turns out to be small, a state of affairs which is, of course, desirable in practice.

3.4.2.3 Impedance Relations

We can now turn to questions of impedance and impedance matching. The constants $\vec{I}_0$ and $\overleftarrow{I}_0$ are related to the corresponding voltage amplitude constants by means of equation 2.2. This equation must be satisfied for all values of x (or of u) when the complete solution 3.22 plus 3.23, together with the corresponding results for the currents, are substituted therein. When second order small terms are neglected it is found that

$$\frac{\vec{V}_0}{\vec{I}_0} = Z_{00} \left[1 + \frac{1}{2pT_0} \frac{d \ln Z_0}{dx} \bigg|_{x=0} \right] \tag{3.24}$$

$$\frac{\overleftarrow{V}_0}{\overleftarrow{I}_0} = - Z_{00} \left[1 - \frac{1}{2pT_0} \frac{d \ln Z_0}{dx} \bigg|_{x=0} \right] \tag{3.25}$$

By making use of the solutions 3.22 and 3.23 separately, in conjunction with the corresponding expressions for the currents, it then appears that

$$\frac{\vec{V}_u}{\vec{I}_u} = Z_0 \left[1 + \frac{1}{2pT} \frac{d \ln Z_0}{dx} \right] \tag{3.26}$$

$$\frac{\overleftarrow{V}_u}{\overleftarrow{I}_u} = - Z_0 \left[1 - \frac{1}{2pT} \frac{d \ln Z_0}{dx} \right] \tag{3.27}$$

---

* The effect is closely analogous to the differentiation produced by a lumped pulse transformer, see § 3.3.2 and equation 1.22.

TRANSFORMERS

It is noticed that the ratio of the voltage and current amplitudes at any particular point differs from the nominal characteristic impedance at that point by an amount which depends on p and therefore on the frequency; further, the ratio is different for the two directions of propagation (compare equation 2.15).

Let us suppose that the line is terminated at the point $x = l$ by an arbitrary impedance Z_2; to find the voltage reflection coefficient we apply OHM's law at the termination and have

$$Z_2 = \frac{\bar{V}_l}{\bar{I}_l} = \frac{\vec{V}_l + \overleftarrow{V}_l}{\vec{I}_l + \overleftarrow{I}_l}$$

Using relations 3.26 and 3.27 we may substitute for the currents in terms of the voltages and obtain the following expression for the reflection coefficient

$$\rho_l \equiv \frac{\overleftarrow{V}_l}{\vec{V}_l} = \frac{\left[1 - \dfrac{1}{2pT_l}\dfrac{d\ln Z_0}{dx}\bigg|_{x=l}\right] Z_2 - Z_{0l}}{\left[1 + \dfrac{1}{2pT_l}\dfrac{d\ln Z_0}{dx}\bigg|_{x=l}\right] Z_2 + Z_{0l}} \qquad (3.28)$$

where second order small quantities are neglected.

It is seen that $\rho_l = 0$, i.e. the termination provides a match, when

$$Z_2 = Z_{0l}\left[1 + \frac{1}{2pT_l}\frac{d\ln Z_0}{dx}\bigg|_{x=l}\right] \qquad (3.29)$$

a value which differs but slightly (at high frequencies) from the nominal characteristic impedance Z_{0l} of the line at the point $x = l$.

When the load consists of a pure resistance of value equal to Z_{0l}, the line then being said to be nominally matched, the expression 3.28 reduces to

$$\rho_l = -\frac{1}{4pT_l}\frac{d\ln Z_0}{dx}\bigg|_{x=l} \qquad (3.30)$$

again to the first order in small quantities.

The input impedance Z_{in} of a length l of line, terminated in an arbitrary impedance Z_2, may now be worked out. By definition

$$Z_{\text{in}} \equiv \frac{\bar{V}_0}{\bar{I}_0} = \frac{\vec{V}_0 + \overleftarrow{V}_0}{\vec{I}_0 + \overleftarrow{I}_0} \qquad (3.31)$$

On substituting for the currents in terms of the voltages by means of relations 3.24 and 3.25 the impedance is expressed in terms of the ratio $\overleftarrow{V}_0/\overrightarrow{V}_0$. By dividing corresponding sides of equations 3.23 and 3.22, and putting $x = l$, this ratio can in turn be written in terms of $\overleftarrow{V}_l/\overrightarrow{V}_l$. This latter ratio is just the reflection coefficient found above and accordingly we have

$$Z_{\text{in}} = Z_{00} \frac{1 + \rho_l e^{-2pU} \left\{ 1 + \left[\frac{1}{2pT}\frac{d \ln Z_0}{dx}\right]_0^l - \frac{1}{4p}\int_0^l \frac{1}{T}\left(\frac{d \ln Z_0}{dx}\right)^2 dx \right\}}{1 - \frac{1}{2pT_0}\frac{d \ln Z_0}{dx}\bigg|_{x=0} - \rho_l e^{-2pU} \left\{ 1 + \frac{1}{2pT_l}\frac{d \ln Z_0}{dx}\bigg|_{x=l} - \frac{1}{4p}\int_0^l \frac{1}{T}\left(\frac{d \ln Z_0}{dx}\right)^2 dx \right\}}$$

(3.32)

where terms in $1/p^2$ have been neglected and the quantity U is the total time delay between $x = 0$ and $x = l$ i.e.

$$U = \int_0^l T(x) dx \qquad (3.33)$$

Since $|\rho_l| \leq 1$ we can expand the denominator by the binomial theorem and obtain

$$Z_{\text{in}} = Z_{00} \left\{ 1 + \frac{1}{2pT_0}\frac{d \ln Z_0}{dx}\bigg|_{x=0} + 2\rho_l e^{-2pU} \left[1 + \frac{1}{2pT_0}\frac{d \ln Z_0}{dx}\bigg|_{x=0} + \frac{1}{2pT_l}\frac{d \ln Z_0}{dx}\bigg|_{x=l} - \frac{1}{4p}\int_0^l \frac{1}{T}\left(\frac{d \ln Z_0}{dx}\right)^2 dx \right] \right\}$$

(3.34)

where terms involving $\rho_l^2 e^{-4pU}$, $\rho_l^3 e^{-6pU}$ etc., which represent multiple reflections up and down the line, have been omitted for simplicity. In the case of the nominally matched line the reduced expression 3.30 for ρ_l is substituted in the result 3.34; as expected, the terms representing multiple reflections now disappear automatically since we are neglecting quantities involving $1/p^2$. The following result is obtained

$$Z_{\text{in}} = Z_{00} \left[1 + \frac{1}{2pT_0} \frac{d \ln Z_0}{dx} \bigg|_{x=0} - \frac{1}{2pT_l} \frac{d \ln Z_0}{dx} \bigg|_{x=l} e^{-2pU} \right] \quad (3.35)$$

FRANK has considered the case of a length of uniform line feeding a tapered line when the characteristic impedance of the former is equal to the nominal characteristic impedance of the tapered line at the junction. The voltage reflection coefficient ρ' observed on the *uniform* line is given by (see relation 2.17)

$$\rho' = \frac{Z_{\text{in}} - Z_{00}}{Z_{\text{in}} + Z_{00}} \quad (3.36)$$

where Z_{in} is the input impedance of the tapered line as found above (equation 3.32). If the tapered line is nominally matched at its output end, substitution for Z_{in} from 3.35 yields

$$\rho' = \frac{1}{4pT_0} \frac{d \ln Z_0}{dx} \bigg|_{x=0} - \frac{1}{4pT_l} \frac{d \ln Z_0}{dx} \bigg|_{x=l} e^{-2pU} \quad (3.37)$$

The reflection coefficient observed on the uniform line provides a criterion for the performance of the transformer which is easily visualized physically.

3.4.2.4 Pulse Distortion

We are now in a position to determine the overall distortion suffered by a short pulse of e.m.f. of amplitude $E(t)$ emanating from a source of internal impedance Z_1. The results will also apply to the case of a source of e.m.f. equal to $2E(t)$, and resistive internal impedance Z_1, which feeds into an arbitrary length of uniform line, of characteristic impedance Z_1, interposed between the source and the transformer.

It will be assumed that the transformer is nominally matched at its output end and multiple reflections will be neglected.

At the transformer input we have

$$\vec{V}_0 = \frac{Z_{\text{in}}}{Z_{\text{in}} + Z_1} \bar{E} \quad (3.38)$$

where

$$Z_{\text{in}} = Z_{00} \left[1 + \frac{1}{2pT_0} \frac{d \ln Z_0}{dx} \bigg|_{x=0} \right] \quad (3.39)$$

for values of the time up to the instant at which the first reflection from the output reaches the input end; equation 3.24, 3.32 and 3.35 all yield this result. On using the solution 3.22 we find for the amplitude of the voltage wave incident at the output

$$\vec{V}_l \propto a \left\{ 1 + \frac{1}{4pT_0} \frac{d \ln Z_0}{dx} \bigg|_{x=0} \cdot \frac{Z_1 - Z_{00}}{Z_1 + Z_{00}} \right.$$
$$\left. + \frac{1}{4pT_l} \frac{d \ln Z_0}{dx} \bigg|_{x=l} - \frac{1}{8p} \int_0^l \frac{1}{T} \left(\frac{d \ln Z_0}{dx} \right)^2 dx \right\} \cdot \bar{E} \quad (3.40)$$

where the factors representing the bulk time delay and the voltage transformation have been omitted; the nominal input attenuation factor a is defined by

$$a \equiv \frac{Z_{00}}{Z_1 + Z_{00}} \quad (3.41)$$

In determining the total output voltage we must take into account the reflection at the termination and write

$$\bar{V}_l = \vec{V}_l + \overleftarrow{V}_l = \vec{V}_l(1 + \rho_l) \quad (3.42)$$

On substituting for ρ_l from relation 3.30 the following result is obtained

$$\bar{V}_l \propto a \left\{ 1 + \frac{1}{4pT_0} \frac{d \ln Z_0}{dx} \bigg|_{x=0} \cdot \frac{Z_1 - Z_{00}}{Z_1 + Z_{00}} - \frac{1}{8p} \int_0^l \frac{1}{T} \left(\frac{d \ln Z_0}{dx} \right)^2 dx \right\} \cdot \bar{E}$$
$$(3.43)$$

Normally the impedance of the source will equal the nominal characteristic impedance of the transformer at the input. This goes without saying when the transformer is used for the purpose of conveying maximum power from the source to the load; the same condition will also usually obtain when we are concerned with a step up in signal amplitude, rather than maximum power transfer, since it will be desirable to match the line nominally at the input end in order to avoid multiple reflections.* However, it is interesting to

---

* When the transformer is fed from a section of uniform line, of internal impedance necessarily equal to its characteristic impedance, we would again arrange that the characteristic impedance of the uniform line was equal to the nominal characteristic impedance at the transformer input.

TRANSFORMERS

note that an output pulse could be obtained, completely free* from distortion provided

$$\frac{1}{T_0} \cdot \frac{d \ln Z_0}{dx}\bigg|_{x=0} \cdot \frac{Z_1 - Z_{00}}{Z_1 + Z_{00}} - \frac{1}{2} \int_0^l \frac{1}{T} \left(\frac{d \ln Z_0}{dx}\right)^2 dx = 0 \quad (3.44)$$

3.4.2.5 Other methods of solution

We are primarily interested in impedances and reflection coefficients rather than in the several voltage and current waves and PIERCE [314] has shown how the simultaneous differential equations in voltage and current can be combined to yield a single first order differential equation in impedance. If $Z = \bar{V}/\bar{I}$ is the impedance looking into the line at the point in question then the following Riccati type equation obtains

$$\frac{Z_0}{p} \cdot \frac{dZ}{du} + Z_0^2 - Z^2 = 0 \quad (3.45)$$

If the reflection coefficient ρ, a function of position, is defined as

$$\rho \equiv (Z - Z_0)/(Z + Z_0) \quad (3.46)$$

then equation 3.45 becomes†

$$\frac{1}{p} \cdot \frac{d\rho}{du} + \frac{1}{2p} \cdot \frac{d \ln Z_0}{du} \cdot (1 - \rho^2) - 2\rho = 0 \quad (3.47)$$

This equation has been treated by WALKER and WAX [315], who use an iterative process and apply the results to the calculation of the resonant wavelengths of tapered lines. BOLINDER [316] assumes that the frequencies are such that $|\rho^2| \ll 1$ i.e. the absolute magnitude of the reflection coefficient is small. It is deduced that the reflection coefficient at the input end $x = 0$ is given by

$$\frac{1}{2} \int_0^l \frac{d \ln Z_0}{dx} \cdot e^{-2\int_0^x j\omega T dx'} dx$$

where it is supposed that the line, of length l, is matched by a resistance equal to its nominal characteristic impedance at the

---

* To the order of accuracy of our present analysis and for values of the time up to that at which the reflection from the output has travelled to the input, been reflected there, and arrived back again at the output.

† The equation is exact when ρ is defined by relation 3.46 but this expression is not equal to the ratio $\overleftarrow{V}/\overrightarrow{V}$ unless the line tapers slowly.

output end $x = l$. On integrating once by parts and neglecting a term containing $\dfrac{d^2 \ln Z_0}{dx^2}$ FRANK's solution 3.37 is obtained.

In BOLINDER's paper a critical comparison is made between the methods here outlined (together with a number of others) as applied to the case of the exponential line for which exact solutions are known. Fourier transform theory of various types of line is also given and examples of design included.

Before passing on to consider a number of lines which have a specific law of impedance variation with position a general method due to CARSON [317] may be mentioned. The equations

$$\left.\begin{aligned}\bar{V} &= \bar{V}_0 - \int_0^x j\omega L \cdot \bar{I} \cdot dx' \\ \bar{I} &= \bar{I}_0 - \int_0^x j\omega C \cdot \bar{V} \cdot dx'\end{aligned}\right\} \quad (3.48)$$

which are simply the formal integrals of equations 2.2 and 2.3, are solved in series form by a process of successive approximation. The method continues to be applicable when finite discontinuities are present and is suitable for numerical computation.

Further information on tapered lines is to be found in papers by ILIN [318], RAYMOND [319] and ZIN [320, 321].

3.4.3 Gaussian Line.

In the ideal case of this particular line the logarithm of the characteristic impedance varies smoothly between $x = -\infty$ and $x = +\infty$ according to a Gaussian law of the form

$$\frac{d \ln Z_0(x)}{dx} = K e^{-h^2 x^2} \quad (3.49)$$

where K and h are constants. In practice, the line must be of finite length, $2l$ say, and the distribution is accordingly only Gaussian over this range (fig. 3.3(a)). Let Z_1 and Z_2 be the impedances between which it is desired to effect a transformation then

$$Z_0(x) = Z_1 \text{ for } x \leq -l$$

and

$$Z_0(x) = Z_2 \text{ for } x \geq +l$$

When these conditions are applied to the integral of equation 3.49 taken between the limits $\pm l$ the constant of integration and the value of K are determined and it follows that

$$\ln \frac{Z_0}{Z_1} = \frac{1}{2} \ln \frac{Z_2}{Z_1} \cdot \left[1 \pm \frac{\mathrm{erf}\,(hx)}{\mathrm{erf}\,(hl)} \right] \quad (3.50)$$

which is the integral Gaussian distribution* for $\ln Z_0$ (see fig. 3.3(b)); the positive sign is taken when $x < 0$ and the negative sign when $x > 0$.

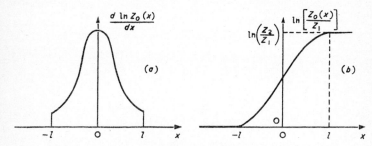

Fig. 3.3. Gaussian line. (Discontinuities in ordinate (a) and in slope (b) are exaggerated.)

We may now rewrite relation 3.49 in the form

$$\frac{d \ln Z_0}{dx} = \frac{h}{\sqrt{\pi}} \cdot \frac{1}{\mathrm{erf}\,(hl)} \cdot \ln \frac{Z_2}{Z_1} \cdot e^{-h^2 x^2} \quad (3.51)$$

and apply FRANK's formula 3.37 to find the voltage reflection coefficient observed on the uniform line $x < -l$ when the output end $x = +l$ is terminated by a resistive load Z_2. It is found that

$$\rho' = \frac{h}{4\sqrt{\pi}} \cdot \frac{e^{-h^2 l^2}}{\mathrm{erf}\,(hl)} \cdot \ln \frac{Z_2}{Z_1} \cdot \frac{1}{p} \left[\frac{1}{T(-l)} - \frac{e^{-2p \int_{-l}^{l} T\,dx}}{T(+l)} \right] \quad (3.52)$$

When designing a line to pass a rectangular pulse with a given degree of distortion we must first estimate the reflection coefficient

---

* The error function is defined by

$$\mathrm{erf}\, y = \frac{2}{\sqrt{\pi}} \cdot \int_0^y e^{-u^2} du$$

that can be tolerated for the lowest frequency component (say $\rho' = 0.1$ at the frequency whose half period is equal to the pulse length). The exponential factor in the last term in the expression 3.52 merely takes account of phase and accordingly we shall be on the safe side as regards variation in amplitude if we take the worst case namely

$$|\rho'| = \frac{h}{4\sqrt{\pi}} \cdot \frac{e^{-h^2l^2}}{\text{erf}(hl)} \cdot \left|\ln\frac{Z_2}{Z_1}\right| \cdot \frac{1}{\omega}\left[\frac{1}{T(-l)} + \frac{1}{T(+l)}\right] \quad (3.53)$$

Before we can choose the quantities h and l we must make sure that the conditions 3.19, upon which the validity of the result 3.37 depends, are satisfied. We first notice that $\ln Z_0$ and all its derivatives are continuous (hence the very low reflections which are found in practice) and *part* of the first condition requires that

$$\frac{h}{4\sqrt{\pi}} \cdot \frac{1}{\text{erf}(hl)} \cdot \left|\ln\frac{Z_2}{Z_1}\right| \cdot \frac{1}{\omega T} \ll 1 \quad (3.54)$$

We have here taken the worst case (the greatest value of $\dfrac{d\ln Z_0}{dx}$ occurs at $x = 0$) and have presumed that T does not vary greatly so that an average value may be substituted. The condition 3.54 is not very stringent in practice and an equality sign may be substituted for a first trial. Further, $\ln(Z_2/Z_1)$ will have an order of magnitude of unity and hence we can write

$$h \simeq 7\omega T \quad (3.55)$$

on the assumption that the values of h and l are going to be such that $\text{erf}(hl) \simeq 1$.

On substituting the equality sign in condition 3.54 we at once find that equation 3.53 becomes

$$|\rho'| \simeq 2e^{-h^2l^2} \quad (3.56)$$

where we have put $T(-l) = T(+l) = T$.

Relations 3.55 and 3.56 determine h and l approximately and the procedure may then be repeated more accurately taking into account the actual values of $\ln(Z_2/Z_1)$ and $\text{erf}(hl)$.

As a check on the validity of the calculations it must be proved that the *complete* first condition 3.19 is satisfied; the integral correction term which appears in expression 3.20 can be worked out to see that its value does in fact turn out to be small compared with unity.

3.4.4 Exponential Line.

Many contributions have been made to the theory and practice of exponential line transformers since a more accurate solution is possible for this type of line. Most writers define an exponential line as one in which the inductance and capacitance per unit length vary with position along the line according to the law

$$L = L_0 \cdot e^{kx}, \quad C = C_0 \cdot e^{-kx} \tag{3.57}$$

where k is a constant either positive or negative.

The nominal characteristic impedance accordingly also varies exponentially

$$Z_0 \equiv \sqrt{\frac{L}{C}} = \sqrt{\frac{L_0}{C_0}} \cdot e^{kx} = Z_{00} \cdot e^{-kx} \tag{3.58}$$

but the time delay per unit length is constant, being given by

$$T \equiv \sqrt{LC} = \sqrt{L_0 C_0} = T_0 \tag{3.59}$$

In the case of rectilinear lines T must necessarily be constant, since it equals the reciprocal of the speed of plane electromagnetic waves in the medium, but in the case of helical lines T may be an arbitrary function of x and independent of the variations in Z_0. A more general type of exponential line may therefore be defined by

$$Z_0 = Z_{00} e^{ku/T_0} \tag{3.60}$$

where once again u is defined by 3.13. When $T = T_0$ everywhere, equation 3.60 reduces to the form 3.58, and u becomes equal to $T_0 x$.

3.4.4.1 Approximate Theoretical Results

Turning now to the initial equations 3.14 and 3.15 and substituting for Z_0 from 3.60 we have

$$\frac{1}{p^2} \frac{d^2 \bar{V}}{du^2} - \frac{k}{p^2 T_0} \frac{d \bar{V}}{du} - \bar{V} = 0 \tag{3.61}$$

$$\frac{1}{p^2} \frac{d^2 \bar{I}}{du^2} + \frac{k}{p^2 T_0} \cdot \frac{d \bar{I}}{du} - \bar{I} = 0 \tag{3.62}$$

It is easily verified that the solution of equation 3.61 is

$$\bar{V}_u = \overrightarrow{\bar{V}_u} + \overleftarrow{\bar{V}_u} \tag{3.63}$$

where

$$\vec{V}_u = \vec{V}_0 e^{u\left(\frac{k}{2} - q\right)/T_0} \tag{3.64}$$

$$\overleftarrow{V}_u = \overleftarrow{V}_0 e^{u\left(\frac{k}{2} + q\right)/T_0} \tag{3.65}$$

$\vec{V}_0$ and $\overleftarrow{V}_0$ are arbitrary constants and q is defined by

$$q \equiv + \sqrt{p^2 T_0^2 + k^2/4} \tag{3.66}$$

Expressions similar to 3.64 and 3.65 are obtained for the currents (equation 3.62), involving two further constants $\vec{I}_0$ and $\overleftarrow{I}_0$; the solutions may be written down from the above by simply changing the sign of k.

We must now examine the significance of the term $\mp qu/T_0$ in the exponent.

(i) At high frequencies when $\omega T_0 \gg |k|/2$ we may, as a first approximation, neglect $k^2/2$ compared with $p^2 T_0^2$ in relation 3.66 and obtain

$$q \simeq pT_0 \tag{3.67}$$

The quantity p then occurs in the results 3.64 and 3.65 only in a factor $e^{\mp pu}$, the solutions accordingly represent waves travelling to the right or to the left respectively with a time delay u between the origin and the point x.

On putting $x = 0 = u$ it is seen that the constants $\vec{V}_0$ and $\overleftarrow{V}_0$ are the amplitudes at the origin of the two voltage waves (thus justifying the notation used). The voltage transformation ratio $\vec{V}_u/\vec{V}_0$, neglecting time delays, for waves travelling to the right from the origin to the point x, given by relations 3.64 and 3.60, turns out to be

$$e^{ku/2T_0} = \sqrt{\frac{Z_{0u}}{Z_{00}}} \tag{3.68}$$

(ii) As a second approximation for q we may put

$$q = pT_0 + k^2/8pT_0 \tag{3.69}$$

Expressions 3.64 and 3.65 then become

$$\vec{V}_u = \vec{V}_0 e^{\frac{ku}{2T_0}} e^{-pu} \left(1 - \frac{k^2 u}{8pT_0^2}\right) \tag{3.70}$$

$$\overleftarrow{V}_u = \overleftarrow{V}_0 e^{\frac{ku}{2T_0}} e^{pu} \left(1 + \frac{k^2 u}{8pT_0^2}\right) \tag{3.71}$$

where terms in $1/p$ of higher order than the first are neglected. These results are identical with those obtained by FRANK (equations 3.22 and 3.23) and accordingly we have simply to substitute the particular exponential law of impedance variation in all the equations of § 3.4.2.3 and § 3.4.2.4 when only approximate results are required.

(iii) The manner in which the transformer treats frequencies $\omega T_0 < k/2$ is better considered purely from the continuous sine

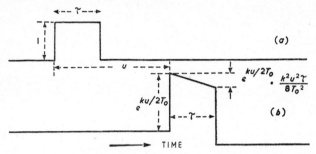

Fig. 3.4. Approximate treatment suffered by a short rectangular pulse on travelling along an exponential line.
(a) Input pulse $x = 0$.
(b) Pulse observed at point x. $u = \int_0^x T(x')dx'$.
Also see Fig. 3.9.

wave point of view. On putting $p = j\omega$ relation 3.66 shows that q is a real quantity. No propagation occurs and the input signal is strongly reflected (the arrows over the symbols now lose their directional significance and merely serve to distinguish one constant from the other). This low frequency cut-off is not complete however, as in the case of a high pass filter or a waveguide of sufficient length, since a d.c. connection exists between the source and the load.

As explained in the penultimate paragraph of § 3.4.2.2 the integrating term in equation 3.70 predicts that the top of an initially rectangular short input pulse will fall linearly by a total fractional amount $k^2 u\tau/8T_0^2$,* where τ is the duration of the pulse, see fig. 3.4

---

* It may be emphasized that this result applies to either of the following cases: (i) the input pulse is obtained from a source of zero internal impedance and the waveform is observed at a point on an infinitely long line or on a perfectly matched line, or (ii) the internal impedance of the source is equal to Z_{00} and the pulse is observed across a nominal matching resistor connected at the end (see equation 3.80).

(accurate graphical results are given in § 3.4.4.3). The following very useful design formula, due to M.H.L. PRYCE, is readily obtained with the aid of relation 3.68

$$\frac{\text{Total delay along line}}{\text{Duration of pulse}} = \frac{50 \, (\ln N)^2}{P} \quad (3.72)$$

where N is the voltage transformation ratio and P is the percentage fall in amplitude at the end of the pulse.

As pointed out in the closing paragraph of § 3.4.2.2 this formula may be used as a criterion for assessing the applicability of the analysis. If we take $P = 10\%$ as a reasonable limit both practically and theoretically it follows that the voltage step-up ratio must not exceed the value 1·5 : 1 for the case of a line in which the delay time is equal to the pulse duration.

For the sake of completeness we shall now list the expressions to which some of the results of § 3.4.2 reduce in the case of an exponential line $\left(\dfrac{d \ln Z_0}{dx} = \dfrac{kT}{T_0}\right)$:—

Voltage/current ratios

$$(3.26) \qquad \frac{\vec{V}_u}{\vec{I}_u} = Z_0 \left[1 + \frac{k}{2pT_0}\right] \qquad (3.73)$$

$$(3.27) \qquad \frac{\overleftarrow{V}_u}{\overleftarrow{I}_u} = - Z_0 \left[1 - \frac{k}{2pT_0}\right] \qquad (3.74)$$

Reflection coefficient at an arbitrary termination Z_2

$$(3.28) \qquad \rho_l = \frac{\left(1 - \dfrac{k}{2pT_0}\right) Z_2 - Z_{0l}}{\left(1 + \dfrac{k}{2pT_0}\right) Z_2 + Z_{0l}} \qquad (3.75)$$

Terminal reflection coefficient when the termination provides a nominal match

$$(3.30) \qquad \rho_l = - \frac{k}{4pT_0} \qquad (3.76)$$

TRANSFORMERS

Input impedance, omitting terms representing multiple reflections

$$(3.34) \quad Z_{\text{in}} = Z_{00}\left\{1 + \frac{k}{2pT_0} + 2\rho_l e^{-2pU}\left[1 + \frac{k}{2pT_0} - \frac{k^2 U}{4pT_0^2}\right]\right\} \quad (3.77)$$

Input impedance of nominally matched line

$$(3.35) \quad Z_{\text{in}} = Z_{00}\left\{1 + \frac{k}{2pT_0}(1 - e^{-2pU})\right\} \quad (3.78)$$

Voltage reflection coefficient observed on a uniform line feeding the transformer which is assumed to be nominally terminated at its output end

$$(3.37) \quad \rho' = \frac{k}{4pT_0}(1 - e^{-2pU}) \quad (3.79)$$

Equation 3.43 shows that the output pulse observed across a nominal resistive termination due to a source of e.m.f. $E(t)$ of internal impedance Z_1 acting at the input $x = u = 0$ is given by

$$(3.43) \quad \bar{V}_l \propto a\left\{1 + \frac{k}{4pT_0} \cdot \frac{Z_1 - Z_{00}}{Z_1 + Z_{00}} - \frac{k^2 U}{8pT_0^2}\right\}\bar{E} \quad (3.80)$$

where the factors representing the bulk time delay and the voltage transformation have been left out, and multiple reflections have been neglected (alternatively, the result can be taken as being quite correct for times up to the instant at which the wave reflected from the termination reaches the output again after having travelled back to the input and been reflected there).

3.4.4.2 General

Information on the exponential line, treated purely from the sine wave point of view, is contained in papers by BURROWS [322]. It is pointed out that the matching into a resistive load, equal in magnitude to the nominal characteristic impedance at the termination, is improved (at the lower frequencies) by connecting an inductance L' in parallel with the high impedance end of the line and a capacitance C' in series with the low impedance end. The magnitudes are determined by the relations

$$L' = 2T_0 Z_0 / k, \quad C' = 2T_0 / k Z_0 \quad (3.81)$$

where the values of Z_0 to be employed are those appropriate to the end in question. The problem of improving the matching by inserting high-pass and modified high-pass filter sections has been carefully investigated by WHEELER [323] and RUHRMANN [324].

In practice the resistive load will inevitably possess a stray capacitance to earth and the usual series inductance may be added to provide compensation for the high frequency components in the pulse (§ 2.4)

Further contributions on the subject of exponential lines are made by MILNOR [325] who analyses a line in which L varies but C remains constant. A lumped tapered line, composed of low-pass filter sections, is also considered; a relatively efficient transfer of energy takes place if the line is made up of a number of symmetrical T-sections, connected in cascade, in which the series inductances increase (for a step-up transformer) and the shunt capacitances decrease in geometrical progression. This law of variation with section serial number simulates an exponential variation of Z_0. Such a line behaves as a band-pass filter; the low frequencies are lost because the line is tapered and the high frequencies vanish because it is lumped. An early paper by WHEELER and MURNAGHAN [326] also contains a description of this kind of transformer.

OGLAND [327] discusses the use of an exponential line for the transformation of rectangular pulses. The transformation ratio which may be obtained with rectilinear lines is small, due to the limited change in impedance which can be brought about even by large changes in the transverse dimensions (see graph, Appendix II). As suggested by FRANK, a tapered piece of high permittivity dielectric material may be inserted to increase the impedance ratio. Helical lines are attractive since a much wider range of impedance can be covered by varying both the pitch of winding and the diameters (see relations 2.56 and 2.57); the physical length is much shorter than that of the corresponding rectilinear line, for the same amount of low frequency distortion, but the high frequency performance is not as good (§ 2.3.3).

In our analysis of tapered lines (§ 3.4.2.2 et. seq.) we have assumed that the nominal characteristic impedance varies smoothly but it can be shown that no such restriction is placed on the values of L and C separately. Construction of a helical line transformer may accordingly be simplified if the diameter and pitch are both changed together suddenly, by comparatively large amounts, at various points along the line, in such a manner that there is no discontinuity in Z_0. Thick wire may be used where the pitch is coarse and a fine gauge employed where the pitch is fine. Discontinuities in the separate values of L and C should be avoided however when good high frequency performance is required.

It is not essential, in practice, to follow the exponential law exactly and construction may again be made easier by winding the helix on an assembly of short conical insulators arranged such that the cone angle changes by a small, but finite, amount from one to the next.

3.4.4.3 Exact Solution

An important contribution to the subject of exponential lines has been made in two papers by SCHATZ and WILLIAMS [328, 329]. The exact inverse Laplace transforms of the complete solution 3.64 plus 3.65 is found when a step function voltage E is applied at time $t = 0$ at the input point $x = 0$. Only voltage step-up transformers are considered explicitly (k positive) and the analysis is restricted to the case of constant delay per unit length ($T = T_0$ everywhere). The solutions are easily generalized however for a variable time delay by replacing x in the results by $\dfrac{1}{T_0} \int\limits_0^x T(x') dx'$.

The expressions for an arbitrary terminating impedance are exceedingly complicated and the solutions are restricted to cases in which the following conditions obtain at the end of the line:

(i) an infinite continuation of the same line,
(ii) open circuit,
(iii) short circuit,
(iv) termination by a resistance equal to the nominal characteristic impedance at the output end.

Even in these particular cases the mathematical expressions are too lengthy to be written out here and only graphical results for the first and last cases will be reproduced.

In figs. 3.5 and 3.6 the voltage and current amplitudes, suitably normalized, are plotted against the parameter $t/T_0 x$ in the case of the infinite line for various values of the *voltage* step-up ratio $e^{kx/2}$. Fig. 3.7 shows the normalized voltage at the end $x = l$ as a function of time when the line is nominally matched by a resistance $Z_{00} \cdot e^{kl}$.

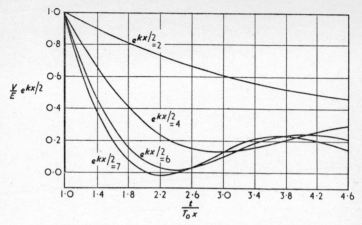

Fig. 3.5. Voltages V at the point x in a step-up exponential line of infinite length due to a step input voltage E at time $t = 0$.

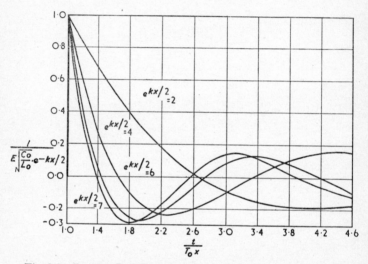

Fig. 3.6. Currents I at the point x in a voltage step-up exponential line of infinite length due to a step input voltage E at time $t = 0$.

The normalized current at the input $x = 0$ is plotted in fig. 3.8 against the parameter $kt/2T_0$ for the infinite line. This curve also applies for any termination up to the time when the first reflection from the output end has travelled back to the input.

The distortion suffered by a short rectangular pulse is illustrated in

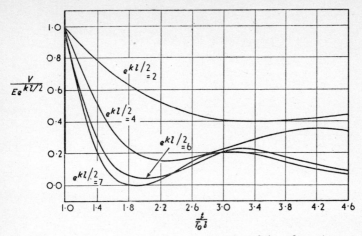

Fig. 3.7. Voltage V at output end, as a function of time, for a step-up exponential line of length l terminated by a load resistance equal to the nominal characteristic impedance at the output.

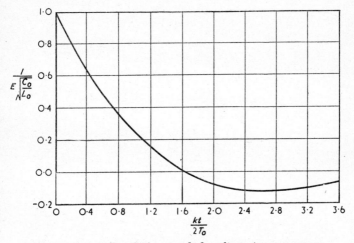

Fig. 3.8. Current I at the input end of a voltage step-up exponential line when excited by a step function voltage E at time $t = 0$. The curve is applicable to the cases of an arbitrary length up to the time of arrival of the first reflection from the output end.

an exaggerated form in fig. 3.9. The voltage and current distributions, on an infinite line, are here sketched for several instants of time at which the pulse has arrived at the points x_1, x_2 etc. The pulse waveform (amplitude versus time), at a particular point x_1 say, will present a very similar appearance but reversed left to right (see fig. 3.4).

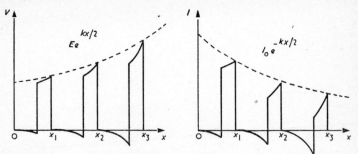

Fig. 3.9. Sketches of voltage and current distribution with position on an infinite line, excited by a short rectangular voltage pulse at $x = 0$, for the several instants of time at which the leading edge of the pulse has reached the positions x_1, x_2, x_3.

Practical design considerations, including the effect of losses in the conductors and in the dielectric material, are also discussed [329]. When the required step-up ratio and the tolerable amount of low

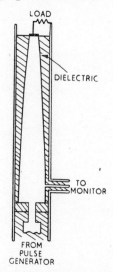

Fig. 3.10. Example of tapered line pulse transformer.

frequency distortion are predetermined the curves enable the rate of taper k and the electrical length to be found. When the transformer is to be used with very short pulses a rectilinear type of line is required, and the average conductor radii must be chosen sufficiently large such that the high frequency distortion due to losses is small.

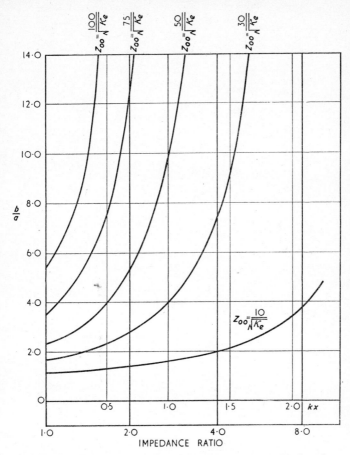

Fig. 3.11. Variation of ratio of conductor radii with length parameter kx, or impedance ratio, for coaxial exponential line pulse transformer.

One must then check that the dielectric losses are not the limiting factor and when only approximate values are required the formulae given in § 2.2.6 may be employed.

The vertical coaxial arrangement of fig. 3.10 is recommended for ease in filling with powder or liquid dielectric. If $a(x)$ and $b(x)$ are the radii of the inner and outer conductors respectively then the formula for the characteristic impedance of a coaxial line (Appendix II) yields the relation

$$\frac{b(x)}{a(x)} = \left(\frac{b_0}{a_0}\right)^{\exp kx} \qquad (3.82)$$

In fig. 3.11 is plotted the ratio of the conductor radii against the length parameter kx for various values of the nominal characteristic impedance Z_{00} at the input.

3.4.5 Linearly Tapered Coaxial Line.

In practice there is not a great deal to choose between the various types of transformer—the Gaussian line gives best performance for a

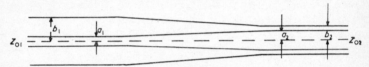

Fig. 3.12. Linearly tapered coaxial line transformer.

given length, the exponential line is most readily dealt with theoretically and a rectilinear coaxial system in which the radii taper linearly is the easiest to construct.

The latter arrangement has been discussed in detail by GENT and WALLIS [330]. We suppose that a uniform coaxial line of given characteristic impedance Z_{01} ohms and given conductor radii a_1, b_1 (see fig. 3.12) is to be transformed into a line of characteristic impedance Z_{02}. It is found that optimum bandwidth is obtained when both conductors are tapered.

(i) If the radii of the second line are at our disposal they should be chosen according to the formulae

$$\frac{a_2}{a_1} = \frac{b_1}{b_2} = e^{(Z_{01}-Z_{02})/120} \qquad (3.83)$$

where $Z_2 = 60 \ln (b_2/a_2)\Omega$ for an air spaced line.

(ii) If one of the radii of the second line, a_2 say, is given, and therefore b_2 also predetermined, we must first taper to values a' and b' determined by relations 3.83 above. A conical section is then inserted to provide a smooth transition from the radii a', b' to the final radii a_2, b_2, all at the constant impedance Z_2 after the manner described in § 2.2.4.2 fig. 2.5(b). The length of the tapered section should be of the same order as the longest wavelength it is desired to transmit but the constant impedance transition need only have a length equal to several times the diameter of the outer conductor at its larger end.

In a specific example, for a 120 Ω to 70 Ω step-down transformer the voltage reflection coefficient viewed in the uniform line, on the high impedance side, was found to be less than $2\frac{1}{2}\%$ at all frequencies for which the taper was more than $1\cdot3\lambda$ long.

3.4.6 Other Laws of Impedance Variation.

BALLANTINE [331] has investigated lines in which both the series impedance and the shunt admittance per unit length are proportional to x^2 or to $1/x^2$ and the analysis has also been extended to analogous lumped lines. STARR [332] has considered the more general case where the impedance and admittance vary as any power x^n and a solution is obtained in terms of Bessel functions. Several particular values of n are discussed as special cases.

ARNOLD and BECHBERGER [333] and ARNOLD and TAYLOR [334] have found the input impedance and attenuation of a line in which the inductance (and series resistance) vary linearly with distance and the capacitance (and shunt conductance) are constant. Such analyses would apply directly to the case of a helical line of variable pitch and constant diameter enclosed in a constant diameter outer sheath. CHRISTIANSEN [335] discusses a four wire open line arrangement which approximates well to an exponential line. The separation between all four wires varies linearly with distance and changes its slope at only one point on the line.

3.5 TRANSMISSION-LINE TYPE OF PULSE INVERTER

The tapered line transformers described above give an output of the same polarity as the input signal. In order to provide a complete alternative to the lumped pulse transformer, with its property of optional phase inversion, it is necessary to have available a transmission line arrangement which will perform the inverting operation.

In the realm of short wave radio transmission considerable attention has been given to the problem of feeding an unbalanced system from a balanced one or vice-versa. This problem, of course, includes that of phase inversion. Networks (BROWN [336]) employing a limited number of lumped reactances have been used in the radio field but the very limited bandwidth of such arrangements prohibits their use in pulse applications. A device known as the "balun" or quarter wave sleeve transformer described, for example, by MARCHAND [337, 338] or FUBINI [339], has been developed by LEWIS and WHITBY [340] for wide band applications.

3.5.1 Principle of Operation.

Consider a length of coaxial cable, of characteristic impedance Z_0, fed at the end A from a source of e.m.f. $2E$ and resistive internal

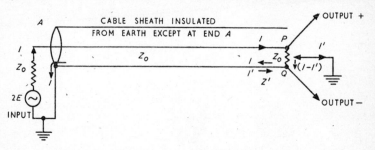

Fig. 3.13. Simple pulse inverter using coaxial cable.

impedance equal to Z_0 (fig. 3.13). The outer conductor of the cable is everywhere insulated from earth except at the input end. A potentiometer of resistance also equal to Z_0 is connected as shown at the far end, and we are interested in the voltage signals, with respect to earth, which appear at the points P and Q.

Once the signal, a voltage step function for example, has been launched into the cable at the end A a voltage wave of amplitude E travels down the cable in the normal principal mode. Clearly the wave propagation is quite unaffected by the position of the potentiometer slider. When the latter is set to the end Q the normal signal E appears at P and no reflection occurs since the potentiometer provides a match. If the slider is now turned to P an inverted output will be observed at Q and if the potentiometer is set halfway a symmetrical output is obtained.

The device clearly ceases to function at low frequencies, when the point P is earthed, since the generator is then short-circuited; performance on long pulses must therefore deteriorate. The voltage wave E has associated with it a current $I = E/Z_0$ which flows along the inner conductor and an equal and opposite current $-I$ which

Fig. 3.14. Ultra-high-frequency choke arrangement.
(*Ruler marked in centimetres.*)

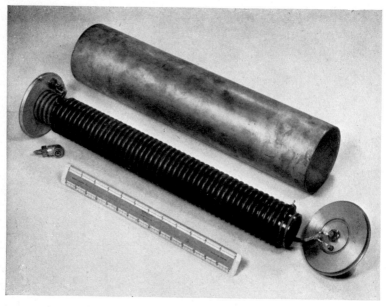

Fig. 3.15. Helical transmission line arrangement.
(*Upper scale of ruler marked in inches.*)

flows in the cable sheath. Because of the skin effect this latter current is confined entirely to the inside surface of the sheath. At the output end a part, I', of the current I flows from the outside of the sheath, round the end of the cable and onto the inner surface of the sheath. It is the difference current $I - I'$ flowing through the terminal resistance which provides the voltage drop across it and gives the negative output signal (the wave reflected at the end Q is neglected for the moment). The current I' is given by the actual negative voltage output at Q divided by the rather indefinite impedance Z' of the outside of the cable sheath viewed from the end Q looking back towards the earthed end A. Owing to the skin effect the currents I and I' are quite independent of one another and our concern is to keep I' as small as possible i.e. Z' as large as possible, over the relevant band of frequencies.

The simplest procedure (fig. 3.14) is to coil up the length of cable following the normal methods of U.H.F. choke construction in which self-capacitance between groups of turns and the capacitance to the earthed screening can, in which the whole arrangement is enclosed, are a minimum. The spacing between the turns of the choke should be greatest at the output end, where the outside of the sheath is live, so that capacitances there are least. The spacing may then decrease towards the input end where the sheath is earthed.

As far as the primary wave is concerned, in the case when P is earthed, the cable appears to be terminated by a resistance Z_0 in parallel with a reactance pL where L is the inductance of the choke. The output pulse, corresponding to a step function input pulse, therefore decays exponentially with a time constant equal to $2L/Z_0$.

In the arrangement shown in fig. 3.14 about 9 yd of 72 Ω cable ($\frac{3}{16}$ in. diameter) are housed in a container $5\frac{1}{2}$ in. long, of diameter equal to $7\frac{1}{2}$ in. A pulse rise time of less than 2 mμsec is obtained with a 12% loss in amplitude. Theoretical and experimental values of 0·8 μsec are obtained for the time constant $2L/Z_0$.

The terminating potentiometer, which we have introduced in order to explain the mode of operation, is not incorporated but two output plugs are provided, one connected to the inner conductor and the other to the sheath, at the output end. A shorting socket is available by means of which the appropriate earth connection to P or Q may be made; an output of either polarity may thus be obtained at will by simply changing over the output lead and the shorting socket.

Advantages possessed by the device include perfect linearity, ease of construction, and the capability of handling high powers. Among disadvantages may be listed (i) considerable physical size, (ii) the time delay suffered in the normal transit down the cable, (iii) the additional, though small, high frequency attenuation due to the normal transit, and (iv) most important of all, limitation to impedances for which flexible co-axial cable is available.

3.5.2 Helical Line Arrangement.

At the higher frequencies with which we are concerned the choke will behave more as a complex transmission line rather than as a pure inductance. An alternative arrangement is shown in fig. 3.15

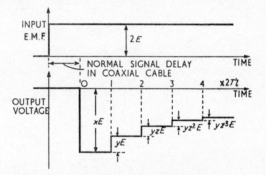

Fig. 3.16. Output voltage at point $Q(P$ earthed) for helical transmission line arrangement.

in which the outside of the cable sheath and the containing cylinder form a transmission line with a helical inner conductor. The normal propagation inside the cable itself is again unaffected by the coiling, and the performance of the inverter at high frequencies should be improved since the stray capacitances to the outer screen now combine with the inductance of the helix to form a line of definite high characteristic impedance Z_0'. Since the latter, which in the case of P earthed appears in parallel with the termination Z_0, is ideally purely resistive, a short input pulse should be inverted with a perfectly flat top which is not exponentially differentiated. The output pulse length, for a step function input, is limited by successive reflections to and fro in the helical line which is shorted at the end A. Each time the wave on the helical line reaches Q most of it is reflected, since $Z_0' \gg Z_0$, but a small voltage change appears across the resistive termination in opposite phase to the main signal. The amplitude of the reflected waves dies away after repeated reflections (even if the line is assumed lossless) and a series of steps of successively decreasing magnitude follow the arrival of the main pulse* (fig. 3.16).

When the reflection at Q in the Z_0 line of the initial pulse is taken into account it is easily shown that the parameters x, y, z of fig. 3.16 are given by

$$x = \frac{Z_0'}{Z_0' - Z_0/2}, \quad y = \frac{Z_0 Z_0'}{(Z_0' + Z_0/2)^2}, \quad z = \frac{Z_0' - Z_0/2}{Z_0' + Z_0/2} \quad (3.84)$$

In this diagram l is the length of the helical line and T'' the delay per unit length in this line. Best performance is obtained when both Z_0' and T'' are greatest i.e. when the inductance is greatest. The outside cylindrical container might well be slotted longitudinally in several places to reduce the shorted turn effect (§ 2.3.1) and the interspace could profitably be filled with a magnetic material.

The high-frequency performance of the device is determined by the variations in Z_0' which occur due to the effects discussed in § 2.3.3. The input impedance of the helical line falls rapidly after the self resonant frequency of one turn is exceeded (equation 2.65).

In the particular arrangement shown in fig. 3.15 about 50 turns of $\frac{3}{16}$ in. diameter 72 Ω coaxial cable are wound on a 2 in. diameter insulated former. The whole is contained in a cylinder 18 in. long and 4 in. diameter; Z_0 thus has a calculated value of about 570 Ω, and a double delay time, in the helical line, of about 40 mμsec is expected. The latter figure was confirmed experimentally and an output pulse fell in steps to one-third of its initial value in a period of

---

* It would be advantageous if the space between the helical line and the outside cylinder were filled with a lossy material, to damp out the reflected waves, provided this could be done without lowering the impedance Z_0' unduly or making its value markedly dependent on frequency at the highest frequencies of interest.

The importance of considering the behaviour of a system to a step function input rather than its response to a train of continuous sine waves is well illustrated here. The input impedance of a length l' of transmission line, short circuited at the far end, varies with frequency between zero, when $l' = n\lambda/2$, and infinity (in the lossless case) when $l' = (2n + 1)\lambda/2$; at frequencies between these resonant points all values of inductive and capacitative reactance are encountered. At first sight this variation of impedance with frequency would indicate a hopeless state of affairs since we have been emphasizing that the input impedance Z' must remain independent of frequency for a distortionless output to be obtained. On the other hand, when a step function input is considered, and the reflections suffered by it are taken into account, the output waveform shown in fig. 3.16 is easily predicted; distortion does in fact occur but its precise nature is resolved quite simply and it is found to be not nearly so bad as might have been feared.

0·3 μsec. A rise time of less than 2 mμsec. was observed and the values $x = 0·9$, $y = 0·1$ and $z = 0·9$ were found to obtain.

The advantage possessed by the helical line arrangement, over the choke method of construction, whereby a short pulse (duration less than $2T'l$) is reproduced with a perfectly flat top, is off-set by the occurrence of small subsidiary pulses after the main pulse. The existence of these is predicted from fig. 3.16 (see § 1.4).

ROCHELLE [341] has developed independently a similar type of inverter which closely resembles the original balun. The coaxial arrangement depicted in fig. 3.13 is simply enclosed in a long

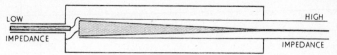

Fig. 3.17. Combined coaxial tapered line transformer and inverter.

cylindrical outer container. The impedance Z' is now the characteristic impedance Z_0 of the rectilinear coaxial line formed by the cable sheath and the outer cylinder. Practical limits on diameter make it impossible to produce an impedance Z'_0 which is very much greater than the impedance Z_0 of the cable and considerable loss in amplitude of the main pulse will result. Again, the double delay time $2T'l$ can only be made long by employing a structure of inconveniently great length. An arrangement 20 ft long is required to pass pulses of duration similar to that handled by the coiled inverters described above. The completely rectilinear system will be superior, however, at the very highest frequencies, since the input impedance at the outside of the cable sheath will remain at the normal value Z'_0 right up to frequencies corresponding to a wavelength of the order of the diameter (3 in.).

A rectilinear inverter and tapered line transformer may readily be combined together as indicated in fig. 3.17. In order that the impedance Z'_0 may be as large as possible compared with Z_0 the live end of the taper sheath should coincide with the low impedance end of the system. The arrangement is, of course, reversible.

3.6 COUPLED LINE TRANSFORMERS

We have seen how lumped pulse transformers and transmission line systems may be employed for the purpose of impedance transformation and phase inversion. Another class of device suggests itself for this purpose which consists of two coupled transmission lines. The primary leakage inductance, and the capacitance of the winding, which were troublesome in the case of the lumped transformer are now spread out into a distributed system. Very good high frequency performance would be expected and the use of ferromagnetic material would be avoided.

FUCHS [342] and BLOCH [343] have given accounts of the theory of coupled open wire rectilinear lines and KARAKASH [344] has used an arrangement such as that depicted in fig. 3.18. This device has

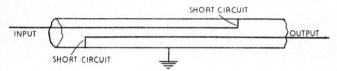

Fig. 3.18. Coupled transmission line type of band-pass filter.

band-pass filter properties but it might be possible to devise a transformer after this fashion, by arranging that the two lines possess a different characteristic impedance.* CROUT [345] has analysed coupled distributed circuits in general terms and applied his method to pulse transformers with distributed parameters.

An analysis has been made by LEWIS [346] of an arrangement consisting of two coupled helices in proximity to an earthed straight conductor.† Such a system can, in general, support waves travelling with two distinct velocities corresponding to in-phase and antiphase modes. An input pulse, at one end of one line, splits into two pulses which travel with the two velocities; both pulses appear on both lines. The device thus, in general, tends to produce double pulses, which are not necessarily each of the same polarity. Steps may be taken to suppress one of them. It is found that the value of the matching resistance required for the terminations is different for the two modes unless the helices are wound in opposite directions, in a concentric system, and the coefficients of inductive and capacitative coupling made equal to one another . In addition to the input and output matching terminations there are two free end points which

---

* The arrangement as it stands will give rise to considerable distortion, but, when the output pulse is to be used simply to trigger some other circuit such distortion may be of no consequence.

† Further information on coupled systems is given by MARSON [347], MATHEWS [348], HUMPHREYS et al. [349].

should in general be correctly terminated if multiple reflections are to be avoided. The incorporation of these extra matching resistances brings about a considerable loss in signal power.

Rectilinear coupled lines would have an advantage in that all waves, in the principal mode, travel with the velocity of light; no disadvantages in the way of double pulse production would therefore be encountered.

RUDENBERG [350, 351] has recently produced transformers in which the primary and secondary windings are wound together as a bifilar coil. Design involves the matching of wave velocities and wave impedances. An unwanted mode may be damped out by placing lossy material in the appropriate region. Bandwidths of up to 500 Mc/s have been measured and no significant resonances were observed. A resolution of a few millimicroseconds was obtained, the upper limit being set by skin effect losses and by the self-resonant frequency (§ 2.3.3.2) of one turn. Such transformers, which contain no ferromagnetic core, can be used for inversion, balance to unbalance conversion, circuit isolation and, of course, voltage transformation.

4

PULSE GENERATORS

4.1 INTRODUCTION

An essential piece of equipment for testing high speed circuits is the pulse generator. A rectangular output pulse is usually required, having as short a rise time as possible, and the amplitude and duration should both be controllable and values known. The various types of circuit readily fall into one of two classes. In the first group the pulses are produced at instants of time determined solely by the generator itself. Usually they occur more or less regularly at a known controllable repetition rate but in some applications a random source of pulses is employed.* In generators of the second class the circuit is triggered by some external agency and we require that the pulse be produced with a minimum of delay, usually, and such delay as there is should not be subject to random variations, i.e. should be jitter free. Circuits falling into the latter class can be employed as primary generators by the addition of a relaxation or other oscillator arranged to produce triggering pulses at the desired recurrence frequency. The rise time requirements on the triggering pulses may be several orders of magnitude less stringent than the requirements which the final output pulse must meet. Conventional microsecond techniques may therefore be employed in this part of the unit.

A circuit described by Espley [402, 403] is of interest in illustrating how far the "conventional" methods of employing several stages of pulse squaring, followed by amplification, may be extended†. The repetition rate is accurately controlled by a 166·7 kc/s quartz crystal and a 35 V 20 mμsec. pulse is obtained across a 35 Ω load. The arrangement, though elaborate, has the advantages of a high recurrence rate and stability in pulse timing to the order of 1 mμsec.

---

* See, for example, a scintillation counter pulse generator described by Wells [401].
† The arrangement is not entirely conventional however as distributed amplifiers (see Chapter 5) are used.

4.2 DISCHARGE LINE TYPE PULSE GENERATORS

The simplest type of pulse generator* is illustrated in fig. 4.1(a). An open ended length of transmission line (or lumped delay line, or even a single condenser) is charged through a high resistance to a D.C. potential V. The line is discharged into a load resistance Z by means of a switching device; this opens after each closure to allow the line to recharge, thus enabling repetitive pulses to be produced. The load Z will normally be equal to the characteristic impedance Z_0 of the line but it is instructive to consider the case in which the

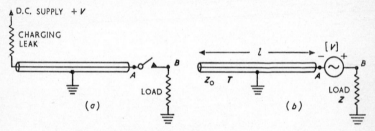

Fig. 4.1. (a) Basic circuit of discharge-line pulse generator.
(b) Equivalent circuit for analysis.

two are different. The effect of closing the switch (see § 1.6) is the same as that due to a generator, connected as shown in fig. 4.1(b), which applies a step function e.m.f. of magnitude V.

On looking into the line at A a certain impedance Z_{in}, given by equation 2.32, is seen. The signal voltage $\bar{V}_B$ which appears across the load Z is given by the simple potential divider formula as

$$\bar{V}_B = \frac{Z}{Z + Z_{\text{in}}} \cdot \bar{V} \tag{4.1}$$

In relation 2.32 ρ_l is put equal to unity since the line is open-circuited at the end remote from the generator, and thus

$$\bar{V}_B = \frac{\dfrac{Z}{Z + Z_0} \cdot (1 - e^{-2pTl})}{1 - \dfrac{Z - Z_0}{Z + Z_0} \cdot e^{-2pTl}} \cdot \bar{V} \tag{4.2}$$

For the case $Z = Z_0$ this reduces to

$$\bar{V}_B = (1 - e^{-2pTl}) \cdot \bar{V}/2 \tag{4.3}$$

---

* See Bibliography—GLASCOE and LEBACQZ, also WHITE [404].

PULSE GENERATORS

and when the inverse Laplace transform is taken it is found that the output pulse is rectangular in shape, has an amplitude $V/2$, and lasts for a time $2Tl$ i.e. the double delay down the line.

In the case when Z differs slightly from Z_0 we may put $Z/Z_0 = 1 + \varepsilon$ where ε is a small quantity. On expanding the denominator of relation 4.2 it is found that

$$\bar{V}_B = \left[\left(1 + \frac{\varepsilon}{2}\right) - e^{-2pTl} - \frac{\varepsilon}{2} \cdot e^{-4pTl} \right] \cdot \frac{\bar{V}}{2} \qquad (4.4)$$

where terms in ε^2 are neglected*.

The output pulse now has the form shown in fig. 4.2. The waveform is drawn for the case $Z > Z_0$ i.e. $\varepsilon > 0$ and further subsequent steps, being of negligible amplitude, are not shown.

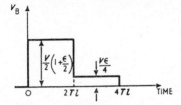

Fig. 4.2. Output pulse when $Z = Z_0(1 + \varepsilon)$, $\varepsilon \ll 1$.

In practice the switch may have a finite, possibly variable, resistance εZ_0 which may be included in the resistance Z for purposes of analysis. The actual output obtained across a load Z_0 will then be equal to $\bar{V}_B/(1 + \varepsilon)$ where $\bar{V}_B$ is as given in equation 4.4 above. The output pulse is still as depicted in fig. 4.2 but the amplitude of the main pulse is now equal to $\dfrac{\bar{V}}{2}\left(1 - \dfrac{\varepsilon}{2}\right)$. An observed experimental

---

* The reader may question the rigour of the present analysis on two points: (i) the switching theorem as stated in § 1.6 applies to a system containing e.m.fs. not necessarily charged capacitors, and (ii) in our Laplace transform analysis we have implicitly assumed that the system is unenergized up to time $t = 0$. The following procedure may be adopted if preferred.

First we note that the Laplace transform of df/dt is $p\bar{f} - f(0)$ where $f(0)$ is the value of the function at $t = 0$. In place of equation 2.3 we then have

$$\frac{d\bar{I}_x}{dx} = -pC\bar{V}_x + CV$$

where V is the d.c. potential to which the line is charged everywhere initially. The wave solutions and the current values are unaffected but a term V/p must be added to the right hand side of relation 2.9. The same relation as before applies at the open end, where no current flows, but the additional term must be included when applying Ohm's law at the load end. Exactly the same final result is obtained.

waveform, like that shown in fig. 4.2, indicates the presence of switch resistance and any jitter in amplitude, particularly in the small secondary pulse, is attributable to resistance variations.

4.2.1 Mechanical Relays.

It is now well established that certain types of electromagnetic relay satisfactorily perform the switching function. Generators employing relay switches naturally belong to the first class mentioned above.

4.2.1.1. Mercury Relays

Simple metal to metal contacts are not usually satisfactory because of contact bounce which may lead to the production of multiple pulses. A technique has been developed by BROWN and POLLARD [405] in which solid platinum contact surfaces are continuously wetted with mercury by means of a capillary connection to a reservoir below the contacts. The mercury film prevents the contact from breaking when the armature bounces slightly on making contact. Only one pulse is accordingly produced for each relay operation. The whole assembly is sealed in an atmosphere of hydrogen at some ten atmospheres pressure. This enables a high voltage gradient to be withstood right up to the instant of contact and the high gas mobility assists in the dissipation of heat. Such relays, which are of the small high speed type, are considered by BROWN and POLLARD to operate satisfactorily if no spark is visible at make i.e. provided the line voltage does not exceed 50 V, or, for higher voltages, provided the energy flowing is not greater than 40 microjoules. It is recommended that the current build up rate should not exceed 25 A/μsec. otherwise the mercury bridge may be vapourized giving a discontinuity which may last for a few microseconds.

Circuits are described by MOODY et al. [406] and by GARWIN [407] using Western Electric relays types 275B, 276B and D–168479. These relays are not coaxial types and require to be mounted in a coaxial housing. Pulse rise (and decay) times of less than 0·2 mμsec. are claimed by GARWIN at repetition rates of 60–120 c/s. It is interesting to note that appreciable distortion is suffered by a pulse on passing through 10 m of solid dielectric shaping cable.

It appears that such mercury relays are reliable, for the present application, at line potentials of less than 10 V (when no spark occurs at make) or at potentials greater than 100 V when a definite gas discharge occurs prior to mechanical make. Two types of contact mechanism may thus be called into play. At intermediate voltages uncertain operation may be encountered in which the main pulse may be reduced in amplitude being followed, throughout an interval of a few microseconds, by irregular secondary pulses.

4.2.1.2 Coaxial Relay

As already mentioned, the above relays were designed as high speed types, with no view to any r.f. application. Their success as

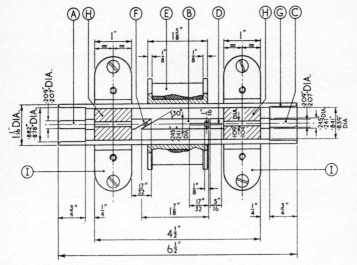

Fig. 4.3. Section through 72 Ω coaxial vibrating switch.

millimicrosecond pulse generators is due to the small size which enables them to be mounted as part of a coaxial system in which little reflection occurs. The design is not ideal however, from the r.f. point of view and WHITBY and LEWIS [408] have constructed a vibrating relay in which attention has been paid to preserving the coaxial nature throughout. Simple platinum contacts are employed and the device, as here described, tends to produce multiple pulses.*

The arrangement, which is comparatively large in size and therefore rugged, easy to construct, and capable of handling higher powers than the mercury relays, is shown in fig. 4.3. The inner conductors A, B, C are made of steel, silver plated to reduce skin

---

* A very short platinum wire dipping into a minute mercury cup might be employed.

effect losses, and the vibrating member B is hinged by means of a piece of steel spring strip D 0·02 in. thick. The magnetic field due to A.C. flowing in the coil E (5000 turns No. 38 S.W.G. enamel), induces opposite magnetic poles across the contact gap F and attraction provides the driving force (in the equilibrium position the member B makes a small angle with the axis and the contacts do not touch). The outer conductor G is split longitudinally into two halves and the whole inner assembly is clamped rigidly against the split perspex insulating supports H by pairs of saddles I ($\frac{1}{8}$ in. thick brass) on the outside of the outer conductor. The longitudinal gap is not quite closed so that the alternating magnetic field shall not be diminished by a solid "shorted turn" effect. The longitudinal gap does not affect the wave propagation in the principal mode. The ratio of the conductor diameters is chosen to give an impedance of 72 Ω and the effect of the dielectric supports H is corrected for, to a first approximation, by the appropriate change in diameter.

The inner and outer conductors are stepped near the ends in order to facilitate plug-in connection into the rest of the circuit. A conical adaptor is required to provide a gradual change from relay dimensions to coaxial cable diameters, as discussed in § 2.2.4.2. A pulse rise time of the order of 1 mμsec. was obtained (the limitation lying in the display equipment used rather than in the relay itself) at a repetition frequency of the order of 100 c/s. At line voltages exceeding 100 V a spark is clearly visible before make. Rise time performance is not impared and the pulse edges are equally sharp at all voltages.

4.2.1.3 Spark Mechanism

It was interesting to observe with the above relay that, at the higher voltages and with a critical amplitude of vibration such that the contacts never touched each other mechanically, the output pulse was as sharp as ever being produced by the gas discharge mechanism alone. Line potentials of 600 V have been used satisfactorily and this suggests that the device, if suitably pressurized to run at much higher voltages, might have useful application as a mechanically triggered high power spark gap modulator [231]. Electrically triggered spark gaps giving very sharp pulses are described by FLETCHER [409], and a surge generator giving 80 KV 1700 A pulses of duration 0·8 μsec. and a rise-time of less than 10 mμsec. has been used by GOODMAN et al. [410].

4.2.1.4 Practical Circuit

A typical relay type pulse generator circuit, for general test purposes, is shown in fig. 4.4. The D.C. charging potential can be easily varied and its value read on a voltmeter. Pulses of either polarity can be obtained at will (a very attractive feature possessed by the relay type of circuit). The relay may be driven from the A.C. mains or from a low frequency oscillator:* the frequency of

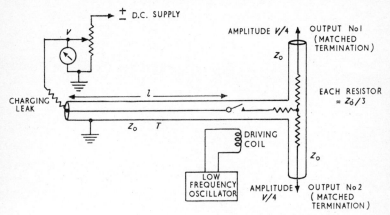

Fig. 4.4. Typical relay type pulse generator giving two outputs.

the latter should be variable so that the relay may be pulsed at the frequency at which it vibrates most smoothly. Assorted lengths of cable, or plug-in sections of coaxial line should be available to provide various pulse lengths. The number of plug and socket connections should be kept to a minimum to avoid unnecessary reflections.

It is important that the charging leak resistor be mounted as closely as possible to the end of the line in order to avoid stray capacitance which would modify the reflection coefficient at the nominally open end. If the relay is prone to give rise to multiple pulses their effect can be much reduced by suitably choosing the cable charging time-constant (equal to the product of the leak resistance times the total capacity of the line). The time-constant should be as long as possible so that negligible partial recharging

---

* When only a very low repetition rate is required the unit may be simplified by using one of the all-relay oscillators described by IVES [411].

can occur between bounces after the first make, but short enough to allow complete charging during the off part of the cycle.

More than one output channel is usually required, and the two-way and three-way matching pads depicted in fig. 4.5 may be used. These provide a resistive match Z_0 looking into any of the arms when all the other arms are terminated by resistances Z_0. Amplitude attenuation figures are quoted. Care must be taken to avoid reflections, due to stray reactance and the resistors must be mounted in some form of coaxial housing (see § 4.5.3).

One output channel is usually required for triggering an oscillograph and the signal pulse will have to be delayed to allow for the

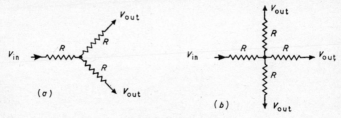

Fig. 4.5. (a) 2-way matching pad $R = Z_0/3$ $V_{\text{out}}/V_{\text{in}} = 1/2$.
(b) 3-way matching pad $R = Z_0/2$ $V_{\text{out}}/V_{\text{in}} = 1/3$.

time-base circuit start-up time. This may be accomplished by including a suitable length of cable* anywhere in the signal circuit under test. When small signal pulses are required, together with a large trigger pulse, an attenuator (§ 4.5) may be inserted in the signal channel.

Such a pulse generator is attractive from the points of view of both simplicity and accuracy in pulse amplitude and duration. The latter depends on the propagation characteristics of the line and it has been found by BROWN [413] that the electrical length of a typical dielectric-filled cable varies by only 0·1% over a period of one year including temperature changes of 10°C.

As an example we may quote from the specification of the Spencer Kennedy Model 503 Pulse Generator. This employs a Westinghouse 475B mercury relay in a 50 ohm coaxial housing. A 2 mμsec. pulse rise time is obtained and various voltage ranges from 0·005 to 100 V are available, the amplitude being variable from zero, in each range,

---

* When a fine control of relative delay between two signal channels is required a telescopic line lengthener, such as the one described by McALISTER [412], may be inserted.

by a potentiometer. Pulses of either polarity may be selected and the recurrence frequency is variable in a number of steps from 45 to 200 c/s.

4.2.1.5 General

A disadvantage of relay type generators is the low repetition rate usually encountered. FINCH [414], however, has used a modified form of the reed type mercury-fed relay, described by BROWN and POLLARD [405], which operates at 480 c/s. A model was later constructed to function at 1000 c/s. These relays are incorporated in an instrument designed primarily for testing networks that can be arranged to store a D.C. charge (a potential of about 50 V is employed). The network is discharged by the relay, and produces a pulse characteristic of itself. The waveform is displayed together with one produced simultaneously by a standard similar network and direct visual comparison can then be made.

While on the topic of relays we may digress to mention certain types, which, although not suitable for pulse generation, have been designed for wide-band r.f. switching purposes. They may be applied to the remote switching of pulse channels but may also be used to advantage in place of conventional wafer or toggle switches even when distant operation is not involved.

The Londex coaxial relay type SCX, employing silver contacts, provides a single pole changeover at impedances of 45 or 72 Ω. The maximum recommended frequency is 150/200 Mc/s and a power of 75 watts can be handled if the line is properly terminated. The capacity between open contacts is of the order of 1 pF.

Messrs. Thomson Products Inc. list a range of 50 Ω impedance single pole multi-way types (including a double-pole changeover unit) which give a maximum voltage standing wave ratio of 1·5 in the range 0–10,000 Mc/s.

4.2.2 Thyratron Pulse Generators

A gas discharge device, or thyratron valve, may be used to perform the switching function in a line type pulse generator. Such tubes can be triggered from an external source of pulses and the arrangement would accordingly be expected to fall into the second class of circuits indicated in § 4.1. The jitter in triggering delay may be too large for some applications however and the generator should rather be treated as belonging to the primary class. Thyratron circuits have one advantage over relay pulse generators in respect of the considerably higher recurrence rate at which they may be operated.

Much work has been done on thyratron design for use in low and high power modulators in the microsecond region. Information may be sought in papers by BIRNBAUM [415], CHANCE [416], KNIGHT [417, 418], MULLIN [419], ROMANOWITZ and DOW [420], WEBSTER [421], WITTENBERG [422], and in the book by GLASCOE and LEBACQZ (see Bibliography). The application of thyratrons to the millimicrosecond range is discussed by YU et al. [423] general conclusions being that a jitter of down to 2 mμsec. can be achieved, but with difficulty, and maximum recurrence frequencies of 5–10 Kc/s are obtainable.

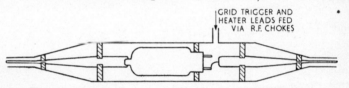

Fig. 4.6. Sketch of coaxial housing for hydrogen thyratron.

The miniature thyratron type 2D21 (CV797) has been used with success. The valve is a xenon filled tetrode capable of passing 500 mA peak current. FOWLER [424] describes its use in a noise source application in which the thyratron discharges a single condenser to produce a saw-tooth shaped pulse which is then differentiated. The ionization time is about 300 mμsec. when the "screening" grid is connected to cathode and a 60 V positive drive pulse is applied to the control grid. The de-ionization time, which determines the time interval which must elapse before the full anode voltage may be restored, can be reduced by holding the control grid highly negative between pulses. Steady operation at a recurrence frequency of 7 Kc/s is obtained when the grid is biassed to -95 V.

The thyratron type 2050 and the hydrogen filled thyratrons types 3C45 and BT79 (see KNIGHT and HOOKER [425], HEINS [426]) have also been employed by other workers with satisfactory results. The hydrogen tubes are suitable for use at recurrence frequencies well in excess of 10 Kc/s. Factors affecting the rise-time, particularly in the case of the miniature hydrogen thyratron type 5C22, are discussed by WOODFORD and WILLIAMS [427]. A rise-time of a few millimicroseconds is obtainable with this valve.

The output pulse rise-time may be limited by circuit inductance

rather than by the thyratron itself. Switching times of down to
2 mμsec. may be obtained when the valve is mounted in a coaxial
housing such as that shown in fig. 4.6. The arrangement, as applied
to a conventional thyratron (even with a top-cap anode), is by no
means perfect from the r.f. point of view. In an ideal design the
anode together with its lead out should be a straight circular rod
and the cathode output should be in the form of a hollow tube of the
same diameter. The emitting surface might be placed across the
end of the cathode tube, the whole thus forming a coaxial line when
enclosed in a cylindrical case. Heater leads and the grid trigger lead
could run inside the cathode tube without affecting the wave
propagation in the main line. Commercial production of such a
thyratron or triggered spark gap would be desirable.

4.2.2.1 Practical Circuit

A practical circuit described by WELLS [227] is shown in fig. 4.7.
The 330 Ω open-circuited cable in the anode circuit of the thyratron

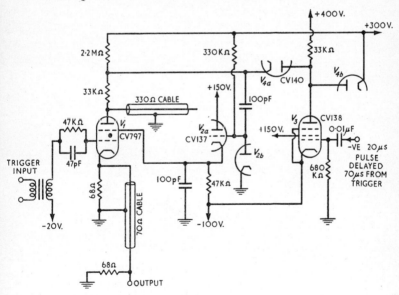

Fig. 4.7. Thyratron discharge-line type pulse generator.

V_1, which determines the pulse length, is discharged into the 70 Ω
output connecting cable. The latter is matched at the near end to
absorb any pulses which would be reflected from the far end should

the load not provide the correct termination. In the circuit as shown the 330 Ω cable is terminated by an impedance equal to that of the thyratron in series with about 35 Ω (the impedance of the output cable in parallel with the 68 Ω resistor). The pulse shaping line is thus not properly matched and multiple pulses are produced by reflection in this line; the secondary pulses which reach the output are of negligible amplitude however.

In order that a well shaped output pulse may be obtained care must be taken in the first instance to reduce as far as possible the stray capacitance to earth at the anode of the valve; the value may then be adjusted to give optimum pulse shape by the addition of a small 5 pF variable capacitor. In the case of the thyratron type 2D21 the screening grid must be taken to earth through a resistance of at least 10 KΩ* by-passed by a small capacitance.

The thyratron grid is biassed beyond cut-off by a potential of − 20 V and the circuit is triggered by applying a large positive pulse. The amplitude of this should be at least 100 V in order to keep the ionization time below 100 mμsec. This triggering delay time was found to be remarkably constant for thyratrons type 2D21 and 2050; random variations did not exceed 1 mμsec. provided the power supplies were stable.

A positive output pulse of 30 V amplitude is delivered into the 70 Ω cable. The amplitude is determined by the supply voltage, the impedance of the pulse shaping cable, and the impedance (35 Ω) at the thyratron cathode plus the series impedance of the valve itself. The latter is small and the output amplitude is almost independent of the thyratron used. Good stability is achieved.

The rise-time† of the front edge of the output pulse is about 10 mμsec. for the 2D21 and about 5 mμsec. in the case of the 2050.

After the thyratron has been triggered the anode voltage must be restored slowly to avoid premature restriking. With the normal anode resistance of about 1 MΩ the recurrence frequency is limited

---

* In the circuit of fig. 4.7 the valve V_2 is cut off by the pulse which appears at the thyratron anode; the input impedance at the cathode accordingly has the high value necessary.

† DIXON and NEHER (unpublished) have found that high voltages of the order of 1–5 KV reduce the conduction time of thyratrons 2D21 and 2050. Immediately after triggering, the current rises to about half the final value in about 10 mμsec. and then increases to the full value in a further short interval of 0·3 mμsec. (in a low inductance circuit). In the case of the 2D21 a peak current of 8 A was obtained in a 50 Ω load.

to about 500 c/s. One method by which this figure may be raised has been mentioned. An alternative arrangement utilizes the fact that the H.T. voltage may be reapplied quickly to the anode of the valve, without causing restriking, provided the anode voltage is kept very low, after each pulse, for a time sufficient to allow the ions to disappear. Recurrence frequencies of up to 10 Kc/s are possible with the circuit of fig. 4.6. After the thyratron has fired the anode potential remains low because of the high 2.2 MΩ resistor. The valve V_3 then drives the anode of the thyratron, through V_{4a}, back to the 300 V supply potential after a delay of 70 μsec. The anode potential is then held at 300 V by the 2.2 MΩ resistor since the diode V_{4a} ceases to conduct after the drive pulse (of duration 20 μsec.) is over.

4.2.2.2 General

A positive output pulse, with a shorter rise-time than that given by the circuit just described, can be obtained by allowing the negative pulse which is available at the thyratron anode to switch off a pentode which is normally taking current. A length of coaxial cable, short-circuited at the far end, is included in the anode circuit of the valve to shape the output pulse.

Two CV2127 pentodes, connected in parallel, each taking about 100 mA of anode current have been used and an output pulse rise-time of about 2 mμsec. obtained. In order that the wattage rating of the valves be not exceeded, and to economize in current, the pentodes are slowly brought into the heavily conducting state a microsecond or so before the output pulse is due, by applying an independent pulse to the grids derived from the trigger pulse. Anode current is then cut off very rapidly by the pulse from the thyratron,* which is applied to the control grids.

The inverting transformers described in § 3.5 are well suited for use with these pulse generators when output pulses, of either sign, are demanded.

---

* A similar sort of "pedestal pulse" technique, used for example by ESPLEY, can often be employed with advantage in cases when a valve stage is required to accept a short negative going input pulse. The screen voltage, rather than the control grid voltage, may be raised comparatively slowly from a low value to the working level a few microseconds before the arrival of the main pulse and afterwards slowly reduced again. When the stage is required to accept positive signal pulses no dissipation difficulties arise since the valve is cut off for most of the repetition period.

4.2.3 Tapered Discharge Line.

It sometimes turns out in practice that the given load resistance and the most convenient shaping line impedance are different in value. A lumped pulse transformer or tapered line is required but we know that both devices produce a fall in amplitude along the top of what should be a rectangular output pulse. An interesting improvement has been suggested by M. G. N. HINE, namely, that the pulse forming line be slightly tapered to a lower impedance at the open end thus generating a pulse with a rising top. This rise may then compensate for the droop inherent in the transformer.

Let us first consider the system depicted in fig. 4.1 for the case when the shaping line is tapered according to an exponential law (see § 3.4.4). We shall suppose that the load impedance Z of fig. 4.1 is a resistance equal to the nominal characteristic impedance Z_{00} of the line at this end; this same end will be taken as the origin $x = 0$ and the positive direction of x is from right to left. On putting $Z_2 \to \infty$ in 3.75 and using 3.77 for the input impedance of the line we have

$$Z_{\text{in}} = Z_{00}[1 + k/2pT_0 + 2e^{-2pU}(1 - k^2U/4pT_0^2)] \quad (4.5)$$

ignoring the effect of multiple reflections up and down the line. It is legitimate to use the approximate expression 3.77, in which terms in $1/p$ of higher order than the first are neglected, provided the desired rise in the output pulse is only to be small. On substituting in equation 4.1 we obtain

$$\bar{V}_B = \left[1 - \frac{k}{4pT_0} - e^{-2pU}\left(1 - \frac{k}{2pT_0} - \frac{k^2U}{4pT_0^2}\right)\right]\frac{\bar{V}}{2} \quad (4.6)$$

For times $t < 2U$, before the major reflection from the open end reaches the load, the output pulse is given by

$$\bar{V}_B = (1 - k/4pT_0)\frac{\bar{V}}{2} \quad (4.7)$$

At time $t = 2U$ the exponential delaying term comes into play and the pulse falls immediately to a value near zero.

When k is negative, that is when the impedance decreases as we move towards the open end, equation 4.7 shows that the output pulse will exhibit a linear rise.

If a lumped pulse transformer, with a differentiating time-constant $T' = L_p/Z_{00}$ (see equations 1.21) is interposed between the line and the actual load to be used the nett voltage across the secondary will be given by

$$\left[1 - \frac{1}{p}\left(\frac{k}{4T_0} + \frac{1}{2T'}\right)\right]\frac{\bar{V}}{2} \qquad (4.8)$$

where the voltage transformation factor has been omitted. Compensation should therefore be achieved if $k = -2T_0/T'$.

Let us now turn to the system shown in fig. 4.8 in which an exponential line transformer BC is connected between the shaping

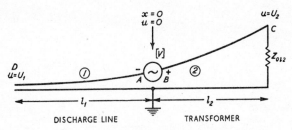

Fig. 4.8. Exponential discharge line in conjunction with an exponential line transformer.

line AD and the load resistance at C which is arranged to equal the nominal characteristic impedance of the transformer at the output end. The signal at B is again determined by equation 4.1 in which Z_{in} refers to the shaping line, having the value given by relation 4.5, and Z is the input impedance of the transformer determined by relation 3.78. We shall continue to neglect multiple reflections in the lines and also assume that the inherent droop in the transformer, and therefore the required compensatory rise, are both small. In place of 4.6 the following expression obtains for $\bar{V}_B$

$$\bar{V}_B = \left[1 - \frac{k_1}{4pT_{01}} - e^{-2pU_1}\left(1 - \frac{k_1}{2pT_{01}} - \frac{k_1^2 U_1}{4pT_{01}^2}\right)\right.$$
$$\left. + \frac{k_2}{4pT_{02}}(1 - e^{-2pU_2})\right]\frac{\bar{V}}{2} \qquad (4.9)$$

where suffixes 1 and 2 refer to the shaping line and the transformer respectively, and we have taken the positive direction of travel as being from right to left in the shaping line and from left to right in the transformer. The exponential term involving the double delay $2U_2$ in the transformer takes into account the wave reflected from the output back to the generator. This contribution can be ignored and we need only consider the outgoing wave $\vec{V}_0 = \vec{V}_B$.

The total voltage $\bar{V}_{l2}$ at the termination C is given by

$$\bar{V}_{l2} = \vec{V}_{l2} + \overleftarrow{V}_{l2} = \vec{V}_{l2}(1 + \rho_{l2}) \tag{4.10}$$

whence, on using relations 3.70, 3.76, and 4.9, we find

$$\bar{V}_{l2} \propto \left[1 - \frac{k_1}{4pT_{01}} - \frac{k_2^2 U_2}{8pT_{02}^2} - e^{-2pU_1} \cdot \right.$$
$$\left. \left(1 - \frac{k_1}{2pT_{01}} - \frac{k_2}{4pT_{02}} - \frac{k_1^2 U_1}{4pT_{01}^2} - \frac{k_2^2 U_2}{8pT_{02}^2}\right)\right] \frac{\bar{V}}{2} \tag{4.11}$$

in which the factors representing the voltage transformation and the time delay along the transformer have been omitted.

The exponential term represents the clipping effect of the shaping line and the extent to which the top of the pulse departs from constant height is determined by the first two integrating terms. These terms cancel if

$$k_1 = -\frac{k_2^2 U_2 T_{01}}{2T_{02}^2} \tag{4.12}$$

It is noted that the impedance of the shaping line should decrease, on looking towards the open end, irrespective of whether a step-up or a step-down transformer is employed. On using relation 3.68 the condition for compensation may be written

$$-\frac{k_1}{k_2} = \ln \sqrt{\frac{Z_{0l2}}{Z_{00}}} = \frac{k_2 U_2}{2T_0} \tag{4.13}$$

for the case when $T_{01} = T_{02} = T_0$.

The transformer must de designed such that a droop of not more than say 10% would occur if a *uniform* shaping line were used. Putting $k_1 = 0$ in relation 4.11 and performing the integration from time $t = 0$ to $t = 2U_1$ we have

$$\frac{k_2^2 U_2 U_1}{4T_0^2} = 0 \cdot 1$$

which, by means of relation 3.68, may be expressed in the form

$$k_2 = \frac{0 \cdot 2T_0}{U_1} \cdot \ln \sqrt{\frac{Z_{0l2}}{Z_{00}}} \tag{4.14}$$

In an actual case the load resistance Z_{0l2}, the impedance Z_{00}, and the pulse length $2U_1$ are predetermined; practical considerations will also enable a value to be assigned to T_0. Relation 4.14 then fixes k_2 and the quantities U_2 and k_1 then follow from relations 4.13. If both the shaping line and the transformer are arranged to have a delay per unit length which is the same everywhere (equal to T_0), then the required lengths of line are given by U_1/T_0 and U_2/T_0 respectively.

A system designed as described gives an output pulse which is sensibly rectangular. The compensation artifice may, of course, be employed to enable a much shorter transformer to be used for a given degree of distortion.

4.3 PULSE GENERATORS EMPLOYING SECONDARY EMISSION VALVES

The most important single contribution to the development of millimicrosecond pulse techniques has been the production of the secondary emission* pentode. Such valves are discussed by ALLEN [429], KEEP [430], MEULLER [431], and VAN OVERBEEK [432], and a grounded-grid triode with one stage of secondary emission is described by DIEMER and JONKER [433]. Tubes of particular relevance are the Philips EFP60, the Osram E2133 and the E.M.I. Type E1945.

A new field of trigger circuits has been opened up and some of these come under the heading of pulse generators. Such circuits are of the regenerative type, employing positive feedback (as in the many flip-flop type circuits familiar in the microsecond range), as opposed to the line discharge pulsers described above. The valves are highly evacuated and the circuits do not involve long triggering delay times, with the consequent possibility of jitter, such as are encountered in thyratron circuits. These generators fall into the second class of § 4.1 and are capable of operating at very much higher repetition rates since there is no dead period corresponding to the de-ionization time of a thyratron.

The EFP60 is essentially a receiving type r.f. pentode which contains an electrode, called the "dynode," placed between the suppressor grid and the anode. The dynode is connected to a source of

---

* A very full bibliography of work on secondary emission phenomena has been compiled by HEALEA [428]. See also the companion volume in the present series by H. BRUINING entitled *Secondary Electron Emission*.

potential, positive with respect to the cathode, and electrons strike it after emerging from the cathode–control-grid–screening-grid–suppressor-grid structure. The dynode surface may be sensitized to emit a copious supply of secondary electrons, on the average about four for each incident primary electron. Nearly all the secondary electrons leaving the dynode are attracted to the anode which is at a still higher positive potential. The valve has the following salient properties:

> (a) The ratio of the mutual conductance (between control-grid and anode) to the total anode capacitance is between two and three times that found in the case of an ordinary r.f. pentode. This improvement is useful in all applications but particularly when the valve is used as a linear amplifier (see Chapter 5).
>
> (b) A nett current flows *out* of the dynode into the external circuit which increases in magnitude when the control grid potential is raised. Signals appearing across a load resistor placed in the dynode circuit are accordingly in phase with the grid signal. By using the dynode as an active electrode, instead of the anode, amplification can be obtained without phase inversion and a regenerative circuit can be constructed by applying feed-back direct from dynode to grid.
>
> (c) Since the anode current is several times the cathode current it is also possible to obtain regeneration by feed-back from anode to cathode—an arrangement which is doomed to failure in the case of the conventional type of valve.

These properties will be brought out more fully in the circuits about to be described.

4.3.1 Simple Trigger Circuit

The simple single-stroke trigger circuit of fig. 4.9 is described by MOODY et al. [406] and operates on the property (b) above. In the quiescent state the valve is biassed off on its control-grid and a positive trigger pulse causes it to conduct. The dynode potential rises and, in so doing, drives the grid positive; the action is cumulative. The output anode also collects current and the anode potential falls until the valve saturates, due to space charge limitation at the dynode. The circuit is of the blocking oscillator type since most of the dynode current flows into the control-grid.

The circuit remains in the meta-stable condition (see waveforms shown in fig. 4.10) until the capacitor from dynode to grid has charged sufficiently such that the grid current falls to a small value.

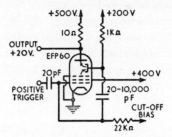

Fig. 4.9. Simple trigger circuit using a secondary emission valve.

The dynode current then begins to fall and its potential drops, this change being again transferred to the grid. The action is cumulative once more; the control-grid is driven negative and the dynode and anode currents are cut off.

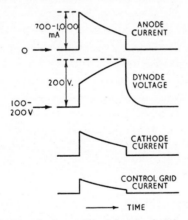

Fig. 4.10. Sketch of waveforms for simple trigger circuit.

The output pulse is taken across a low resistance in the anode circuit and has a rise-time of about 2 mμsec. A peak anode current of up to 1 A is obtainable. Disadvantages of this simple circuit are that the output pulse amplitude and duration are not very well defined, and, that a grid signal of some 6 V amplitude is required for triggering. A more sensitive circuit of improved performance is described in § 7.5.

4.3.2 Practical Circuit

At the suggestion of G. A. HOWELLS, the third property of a secondary emission pentode enumerated in § 4.3 has been applied in a pulse generator developed by WELLS [434] at A.E.R.E. Harwell. This method of deriving positive feed-back has the advantage that the control grid is left free for the application of a (positive) triggering pulse from a comparatively high impedance source. The circuit is shown in fig. 4.11 and delivers an output pulse of up to 10 V amplitude, of either polarity, into a matched 100 Ω cable with a rise-time of 8 mµsec. A triggering signal of some 5 V amplitude is required.

The short-circuited delay line in the anode circuit of V_1 determines the pulse length. A duration of 50 mµsec. is obtained with the values shown and longer pulses can easily be obtained by switching in additional sections. The feed-back capacitor C_1 must be chosen such that, with the delay line terminated with its characteristic impedance of 330 Ω (instead of being short-circuited), the pulse duration (which is then only controlled by C_1 and the circuit resistance) is about 2–3 times the required value. The germanium crystal diode X_1 is included to pass the large grid current which flows after the circuit is triggered.

The output from V_1 is taken from the dynode as a positive pulse. Its amplitude is somewhat indeterminate and the pulse accordingly requires to be limited. In the quiescent state the valve V_1 is cut off and the dynode rests at earth potential via the 3·9 KΩ resistor. No signal reaches the grid of V_2 until the dynode pulse has risen sufficiently to overcome the bias on the crystal diode X_2; the top of the pulse is then limited by the diodes X_3 and X_4. The pulse size can be varied by means of the potentiometer R_2 which determines the bias on X_2.

When germanium crystals are used in this manner as voltage limiters there exists a time delay, of the order of 10 mµsec., between the application of the voltage pulse and the passage of the full forward current through the crystal. The forward resistance of such crystals appears to change in a truly exponential manner, with a time constant of 10 mµsec., from the initial high value to a final value corresponding to the applied voltage. The limiting effect of the crystals X_3 and X_4 is therefore apparently delayed and overshooting of the leading edge of the dynode pulse would occur. This may be compensated by the addition of the resistance R_1 shunted

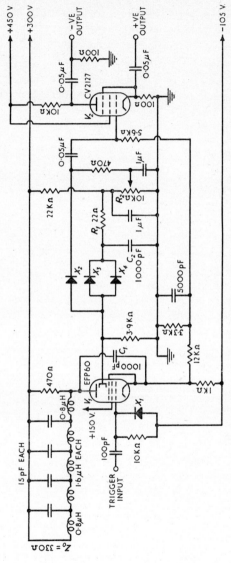

Fig. 4.11. Triggered pulse generator employing one secondary emission valve. (From *Nucleonics*, Volume 10, No. 4, pages 28–30. Copyright 1952, McGraw-Hill Publishing Company Inc.)

by the capacitor C_2 shown in fig. 4.11. The time constant C_2R_1 is chosen to give optimum shape to the final output pulse from V_2.

The CV2127 output valve V_2 is normally biassed off and then caused to conduct by the dynode pulse from V_1. A positive-going output pulse is obtained from the cathode and a negative one from the anode.

4.4 FURTHER POSSIBILITIES

Some other devices, for use as pulse generators, are worthy of mention.

HASTED [435] has used a specially constructed tube, similar to a klystron, in which the second resonator (or catcher) is replaced by a non-resonant coaxial line collector. A beam of electrons from a cathode gun is velocity modulated by a 210 Mc/s sinusoidal voltage applied across two grids. Bunching occurs in a drift space some 40 cm long and pulses are produced at the collector. A pulse length of 0·2 mμsec. is estimated theoretically.

A tube 6BN6, described by ADLER [436] and HAASE [437], has been developed by the Zenith Radio Corporation for use as an improved limiter-discriminator for frequency modulation and television applications. The valve is a receiving type employing a sharply focussed electron beam which is gated by a potential applied to the control grid. Good performance as a squaring device is claimed for all frequencies of sinusoidal input up to 30 Mc/s.

MILLER and MACLEAN [438] describe a new miniature non-gaseous trigger tube type NU-1032-J. Electrons from a triode input section impinge on a dynode. The secondary electrons ejected are collected on two output electrodes which may be used separately or as a unit. An emissive surface having long life and good stability is described and various circuit applications are suggested (applying rather to the microsecond range). The idea of constructing a complete trigger circuit in one envelope, thus reducing stray reactances to an absolute minimum, is attractive, and the device might profitably be developed for use in very high speed circuits.

The performance of blocking oscillators has been investigated in the microsecond range (see, for example, a paper by BENJAMIN [439]) and O'DELL [440] (see § 4.8) describes a circuit which will trigger on a pulse of amplitude 1 V and duration 10 mμsec. A useful pulse generating circuit might well be constructed employing the transformer design techniques given in § 3.3 and using an EFP60 type of valve.

Pulseactor circuits, described for example by HUSSEY [441] and MELVILLE [442], have been used to produce pulses well below 100 mμsec. in duration at frequencies up to a few megacycles per second.

4.5 ATTENUATORS

The attenuator, an arrangement employed solely for the purpose of reducing the amplitude of a pulse by a known amount, is one of the basic circuit "bricks" which may be found, in some form or other, in most pieces of equipment. It is one of the few devices which, if properly constructed, is capable of performing with high accuracy and finds its place among the test-gear in any laboratory. Design is not straightforward, when operation in the millimicrosecond range of frequencies is desired, and the development work done in the field of short-wave radio finds application here. Progress, with increase in bandwidth, has taken place in the following stages:

(a) Simple open wire types employing conventional lumped resistors.

(b) Lumped resistors contained in cylindrical housings and disc forms mounted in coaxial lines.

(c) Lossy transmission lines.

These several types will now be discussed.

4.5.1 Limitations of Resistors at High Frequencies.

The conventional attenuator, illustrated in its simplest possible form in fig. 4.12, employs ordinary resistors with open wiring

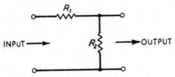

Fig. 4.12. Simple attenuating section.

between elements and for the input and output connections. Performance is usually limited to below 10 Mc/s by stray reactances (series inductance and shunt capacitance) in the wiring, but the simple types are worthy of mention since, being composed of standard components, they are constructed very much more readily than the superior types described later.

The resistors themselves have frequency dependent properties as discussed by BOELLA [443], HOWE [444, 445], PAVLASCK and HOWES [446],* PUCKLE [447], and SIMMONDS [448]; also see Bibliography—BLACKBURN. Resistors may be either of the wire-wound type (which are generally unsatisfactory owing to the inductance present in the winding) or of the pyrolytic carbon or composition types (see PLANER and PLANER [450]). In the pyrolytic form a carbon film is deposited by heat treatment on a ceramic rod. A spiral groove is then cut in the coating to raise the value of the resistance but an increase in inductance is thereby brought about. The composition variety is made of a solid rod composed of various grades of carbon, together with a filler and a resin binder. This last type is preferable from the point of view of low inductance but instability in resistance value may be encountered.† It is known that the resistance falls at high frequencies (the Boella effect) due, apparently, to inherent self capacitance effects. As might be expected the effects are greater for the higher values of resistor and the series resistor R_1 of fig. 4.12 should not exceed 1 KΩ in value. The resistor R_2 should not be too small (say $> 10\ \Omega$) otherwise circuit inductance will give rise to frequency dependence. These values are extremes and an attenuating factor of not more than ten times should be sought after in a single stage.

4.5.2 Simple Lumped Attenuators.

The familiar T-type attenuating pad is shown in fig. 4.13(b). The arrangement has a definite resistive characteristic impedance and may therefore be inserted in a length of transmission line and any number of sections may be connected in cascade to give any required degree of attenuation. DAWES et al. [453] give an account of attenuating networks when stray reactances are taken into consideration. The results, listed in fig. 4.13, are deduced by a Laplace transform

---

* A further report by PAVLASCK and HOWES [449] calculates the effect of frequency variation on the power dissipated when rectangular pulses are applied.

† GRISDALE et al. [451] describe boro-carbon film type resistors which have high stability and a small and predictable temperature coefficient (actually less than that obtaining in the case of wire-wound resistors); low noise properties are also possessed. The construction of metal film type resistors is discussed by HERITAGE [452].

PULSE GENERATORS

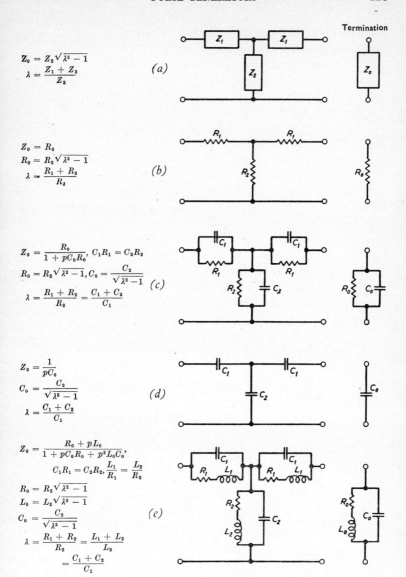

$Z_0 = Z_2 \sqrt{\lambda^2 - 1}$
$\lambda = \dfrac{Z_1 + Z_2}{Z_2}$ (a)

$Z_0 = R_0$
$R_0 = R_2 \sqrt{\lambda^2 - 1}$
$\lambda = \dfrac{R_1 + R_2}{R_2}$ (b)

$Z_0 = \dfrac{R_0}{1 + pC_0R_0}, \; C_1R_1 = C_2R_2$
$R_0 = R_2\sqrt{\lambda^2-1}, C_0 = \dfrac{C_2}{\sqrt{\lambda^2-1}}$
$\lambda = \dfrac{R_1 + R_2}{R_2} = \dfrac{C_1 + C_2}{C_1}$ (c)

$Z_0 = \dfrac{1}{pC_0}$
$C_0 = \dfrac{C_2}{\sqrt{\lambda^2 - 1}}$
$\lambda = \dfrac{C_1 + C_2}{C_1}$ (d)

$Z_0 = \dfrac{R_0 + pL_0}{1 + pC_0R_0 + p^2L_0C_0}$,
$\quad C_1R_1 = C_2R_2, \dfrac{L_1}{R_1} = \dfrac{L_2}{R_2}$
$R_0 = R_2\sqrt{\lambda^2 - 1}$
$L_0 = L_2\sqrt{\lambda^2 - 1}$
$C_0 = \dfrac{C_2}{\sqrt{\lambda^2 - 1}}$ (e)
$\lambda = \dfrac{R_1 + R_2}{R_2} = \dfrac{L_1 + L_2}{L_2}$
$\quad = \dfrac{C_1 + C_2}{C_1}$

Fig. 4.13. Attenuating T-sections. Voltage attenuation per section = $\lambda - \sqrt{\lambda^2 - 1}$ in all cases when properly terminated.

method. It is shown that distortionless attenuation is obtained provided $(Z_1 + Z_2)/Z_2$, (see fig. 4.13(a)), is independent of p, i.e. is frequency invariant. The circuits (c), (d) and (e) are of particular interest when the load has a considerable input capacitance; the condensers C_1 and C_2 are deliberately added across the resistance elements in order to compensate for the capacitance of the termination.

4.5.3 High Frequency Improvements.

Improved high-frequency performance is obtained when the resistors are mounted in closely fitting cylindrical metal cans, as used by GARWIN [407], FLETCHER [409] and others, to reduce

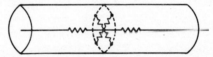

Fig. 4.14. T-section attenuator in coaxial housing.

circuit inductance.* The series arms R_1 conform naturally to such an arrangement and satisfactory resistors can be made of carbon deposited on insulated rods (VAN ROOSBROCK [454]). Standards which function up 1000 Mc/s have been developed. The shunt element R_2, in the simple lumped T-section, does not lend itself immediately to cylindrical symmetry and four lumped rods, as indicated in fig. 4.14, may be employed up to 100 Mc/s. A further step is to substitute a disc of resistive material or to employ an insulated disc coated with carbon. Surface film type resistors will clearly be preferable to solid types if the thickness of the film is less than the skin depth at the highest frequency involved; the variation of resistance with frequency, due to skin effects, will then be absent.

Details are given by ELLIOTT (455) of the attenuating unit shown in fig. 4.15. The series arms are carbon coated ceramic rods and the shunt ceramic disc is coated on one side only (deposition of both sides introduces excessive parasitic reactance). The attenuator, as shown, is fitted with standard plug ends. Messrs Leeds and Northrup have produced an attenuator in which each section is a plug-in unit in the shape of a "U". Several units are mounted on a circular drum and a plugging action is required to effect a change from one section to another. These attenuators have been made to cover all

---

* GARWIN finds that Allen-Bradley 1/2 watt carbon resistors have good characteristics for rise-times of about 0·2 mμsec.

loss ranges from 1 to 50 db at characteristic impedances of 50 or 75 Ω. The change in loss with frequency is less than 1 db up to 1000 Mc/s.

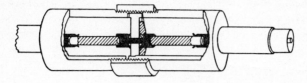

Fig. 4.15. Plug-in coaxial T-attenuating unit.

A rotating turret assembly employing a wiping switch action rather than a discrete plug-in movement gives easier manual operation but lower performance; loss changes of 0·1 db from 0–200 Mc/s and of 1 db from 0–500 Mc/s were found to occur.

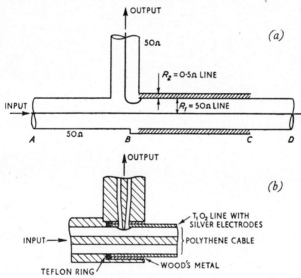

Fig. 4.16. (a) Transmission line potential divider.
(b) Probe connection.

High performance attenuators have been designed by FLETCHER [409] in which the resistive characteristic impedance possessed by a transmission line is turned to advantage. An attenuation of 100 : 1 has been obtained, in one step, with the arrangement of fig. 4.16(a) which is derived from the simple attenuator shown in fig. 4.12 by arranging that the input impedances of two concentrically disposed coaxial lines form the series and shunt arms R_1 and R_2. The low impedance line consists of a thin walled extruded TiO_2 ceramic tube with silver electrodes deposited on both the inner and outer surfaces. The commercial material has a relative permittivity of about 85 which is independent of frequency up to 30,000 Mc/s.

Reflections occur in the input line at B and also at the ends C, D of the lines BC and BD. If the transit times AB, BC and BD are greater than half the pulse duration, reflected signals will not appear at the output until after the pulse is over. If the source A and the ends C and D were perfectly matched, the lengths would, of course, be immaterial; this suggestion, however, merely raises the problem of how to provide accurate matches.

It was found that a "spike" occurred at the beginning of the output pulse waveform due to the excitation of higher modes (§ 2.2.4.2) in the concentric coaxial lines near the point B. These modes die out completely within a short distance down the line and the output lead is accordingly arranged to pass through a small hole in the ceramic tube and connect with the inner silver coating at a point inside the line BD. The assembly is shown in fig. 4.16(b) and it is estimated that less than 10% distortion will be suffered by pulses with a considerably shorter rise-time than the value of 0·4 mμsec. used in the tests.

4.5.4 Lossy Transmission Line Types.

A further stage in development is to regard the attenuator as a lossy transmission line, with distributed circuit parameters. KOHN [456], and BURKHARDTSMAIER [457] have investigated the effect of the coaxial housing on a single rod type resistor shorted to the case at the far end (such as might be used to provide a matched termination to a length of coaxial cable). It is found that the resistor can be of a considerable length and is thus capable of dissipating a large amount of power. The input impedance is essentially resistive when the characteristic impedance of the concentric system is properly chosen. The arrangement is particularly effective when used in conjunction with a compensating circuit the dimensions of which are also deduced.

CARLIN [458] investigates the properties possessed by an infinite length, and by a short-circuited finite length, of lossy uniform transmission line. The short-circuited length provides a useful circuit element, for shunt connection, which may be applied to the compensation of mismatches present in a transmission system. The arrangement has been applied to the "chimney" attenuator sketched in fig. 4.17.

This attenuator is composed of a principal resistive element A, consisting of metallized glass, the resistance of which is not subject to variation over the frequency band (see § 2.2.6.2). The section can be properly matched by the addition of the two lossy shorted stubs

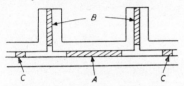

Fig. 4.17. Chimney attenuator.

B together with the short series sections C. All the sections A, B and C contribute to the loss and a 10 db attenuator can be constructed to function from 0–1000 Mc/s (or from 1000–4000 Mc/s). The reflection coefficient is not zero, actually being about 10%, but it is substantially constant over the frequency band. Details of design are given in the paper.

4.6 REFLEX PEAK VALVE-VOLTMETER

Having surveyed a number of pulse generating circuits we now turn our attention to pulse measuring equipments. The cathode ray oscilloscope is by far the most versatile test instrument and Chapter 6 is devoted to its study. When routine monitoring of pulse amplitude is required, however, a valve-voltmeter provides a comparatively simple, inexpensive and accurate means of measurement.

O'Dell [440] has developed a valve-voltmeter which measures the peak amplitude of pulses having a duration of 10 mμsec. or longer.* The instrument gives a direct reading and full-scale deflection on the output meter corresponds to an input of 100 V. A change of only 0·2% full scale reading occurred when the mains input voltage was varied by 25%, and, with the particular component values used,

---

* Details of the Reflex Peak Voltmeter are published by permission of the Admiralty, this equipment being the subject of an Admiralty Patent Application.

a change of less than 0·5% was observed when the input repetition rate was altered from 200 c/s to 7 Kc/s (a 1 μsec. pulse was used for the latter test in order to facilitate comparison with other methods of measurement).

4.6.1 Principle of Operation.

The circuit is as shown in fig. 4.18 and operates on the familiar slide-back principle. A blocking oscillator V_2 is used as the indicator of equality between the input pulse peak and the reference voltage, and adjustment of the reference or slide-back voltage is brought about automatically. The valve V_1 is included to amplify the peak portion of the input pulse and to prevent the blocking oscillator pulse from appearing at the input terminal. A tertiary winding on the blocking oscillator transformer feeds a detector valve V_3 which also acts as an amplifier with load resistance in its cathode circuit. The negative-going output from V_3 is smoothed and fed back, as a D.C. bias, to the control grid of the buffer valve V_1 via the resistor R_2. A robust voltmeter VM displays the value of this bias voltage.

On applying a positive-going pulse train to the input terminals, the blocking oscillator V_2 is triggered via V_1 on every pulse, until enough bias has been developed at the cathode of V_3 (and hence the grid of V_1 driven sufficiently negative) to prevent the further firing of V_2. Intermittent triggering of V_2 then occurs, so that the feed-back voltage is maintained approximately equal to the peak amplitude of the incoming pulses (assumed to be all of the same amplitude but not necessarily regular in time). The blocking oscillator triggers on a pulse amplitude of about 1 V and the effect of this difference between input amplitude and slide-back voltage can be removed by suitable adjustment of the zero of VM. Fluctuations in triggering level will be much less than 1 V and accordingly good accuracy is obtained on 100 V input pulses.

The 100 Ω carbon resistor R_1 is included in the input to damp the oscillatory circuit composed of the stray inductance of the blocking condenser C_1, and its leads, and the input capacitance of V_1. The actual value required may vary in individual instruments but some damping is essential when very short pulses are to be measured. The change with repetition rate quoted above is probably due to the loss of D.C. level in the input circuit owing to the presence of C_1. Provision might be made for D.C. restoration when the pulse mark to space ratio ceases to be very small.

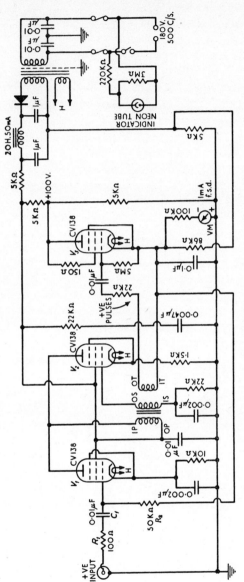

Fig. 4.18. Reflex peak valve-voltmeter (after O'Dell).

The instrument is rather sensitive to the presence of stray pulsating electromagnetic fields and is enclosed in a thick aluminium case. Capacitors are connected between each mains lead and earth to prevent spurious triggering of the blocking oscillator by mains borne interference.

4.6.2 Blocking Oscillator Transformer Construction.

Details of the blocking oscillator transformer are shown in fig. 4.19(a). It is designed to give a relatively long pulse, about 8 μsec.,

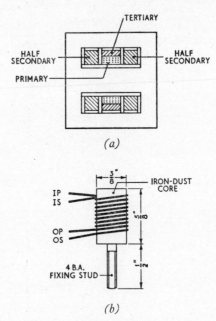

Fig. 4.19. (a) Main (b) auxiliary blocking oscillator transformers (after O'DELL).

and the grid circuit time-constant was chosen so that recovery was rapid compared with the minimum expected time interval between pulses. The transformer core consists of a $\frac{1}{2}$ in. thick stack of Radiometal stampings type 470T and the coils are wound on a $\frac{1}{2}$ in. (internal) square section tufnol former ($\frac{1}{16}$ in. thick) with $\frac{1}{16}$ in. thick tufnol end cheeks. The primary is wound with 60 turns of No. 32 S.W.G., D.S.C. wire and the secondary, in two halves, consists of 100 turns, each half, of similar wire. The tertiary winding is made up of 100 turns of No. 36 S.W.G., D.S.C. wire. Every layer is insulated with paper, 0·002 in. thick, and two layers of 0·01 in. thick empire cloth are interposed between the primary and tertiary windings.

When pulses of duration 10 mμsec., or less, are to be measured the transformer is inoperative, due to the effect of its winding capacitances. A small auxiliary transformer, illustrated in fig. 4.19(b), is accordingly employed. The primary and secondary windings of the auxiliary transformer are connected in series with those of the main transformer (between the latter and the valve). The regenerative action is initiated by the small transformer and continued and maintained by the main transformer. The primary and secondary windings, of the auxiliary transformer, are both of No. 30 S.W.G., D.S.C. wire wound 10 turns per inch directly on a dust core (which has a relative permeability of about 12). The secondary turns are wound centrally between those of the primary, and the wire is fixed to the core by distrene cement.

A wide variety of methods of measuring high fequency voltages are reviewed by SELBY [459]. The majority of the arrangements discussed however are applicable to sinusoidally varying signals and not to pulses.

5

AMPLIFIERS

5.1 INTRODUCTION

THE problem of the amplification of pulses, without distortion of waveform, is a fundamental one in nearly every branch of applied electronics. A great deal of effort has been expended in the design of amplifiers for use in the border line region between the microsecond and millimicrosecond ranges and it would be invidious to select the work of one or two particular individuals by way of introductory references here. The advent of certain high-slope pentodes, with very close spacing between cathode and control grid, and of the secondary emission valves mentioned in § 4.3 enables familiar high gain voltage amplifier circuits to be employed well into our present range of interest.

In order to appreciate the difficulties which face us in the millimicrosecond region, both in valve construction and in circuit design, we must be reminded of the high-frequency limitations to which conventional valves and conventional circuits are subject.

5.2 PROPERTIES OF VALVES

The simplest possible amplifying stage, of the resistance-capacitance coupled type, is shown in fig. 5.1. When a pentode valve is operated with a constant screen-cathode potential the output voltage is $g_m R$ times the input voltage. The actual input voltage at the grid depends, of course, on the impedance of the source and the input impedance (loading effect) of the valve; at low frequencies the valve input impedance is taken to be infinite. The effective value of the anode load resistance R includes the input impedance of the circuit which the amplifier is feeding. At high frequencies various shunt impedances become of importance and limit the stage gain by lowering the effective value of the anode load.

The following effects, discussed for example by EMMS [501] must be borne in mind in valve and circuit design.

5.2.1 High Frequency Limitations.

5.2.1.1 Input Capacitance

A pentode valve possesses a certain capacitance between the control-grid and earth which is composed of the capacitance to the cathode, to the screening-grid, and to the external shield (if any) in decreasing order of importance. Ordinary electrostatic values obtain when the valve heater is cold, but under normal operating conditions

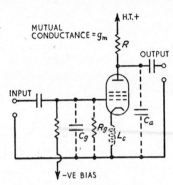

Fig. 5.1. Resistance coupled amplifier stage showing stray impedances.

there is an increase of between 1 and $2\frac{1}{2}$ pF in the grid-cathode capacitance, for most valves (depending on the grid bias); the increase is due to an effective reduction in spacing caused by the presence of space charge. The stray capacitance to earth of the wiring, usually amounting to some 5 pF, must also be added; the total is equivalent to the capacitor C_g of fig. 5.1.

The effect of grid-cathode capacitance can be reduced by inserting a small resistance between cathode and earth. The resistance introduces some negative feed-back, and reduces the gain of the valve, but the fall in capacitance and the consequent lighter loading on the previous circuit may more than compensate for this.

5.2.1.2 Output Capacitance

The anode has a resultant capacitance C_a to earth made up of that between the anode and (i) the external shield* (if any), (ii) the internal

---

* If instability troubles are not likely to occur such screens may be removed from valves which are normally fitted with them; appreciable reduction in capacitance results.

supports, (iii) the supressor-grid and (iv) the screening grid (in descending order of magnitude). The stray capacitance to earth of the wiring must again be added.

5.2.1.3 Input Conductance

At the higher frequencies an effective shunt resistance appears between the grid of the valve and earth. Dielectric loss in the valve holder and valve base may contribute to this loading of the input circuit but it is mainly due to (a) feed-back loss and (b) transit-time loss. The equivalent resistance R_g is proportional to the reciprocal of the square of the frequency.

(a) Feed-back loss is caused by the stray cathode inductance L_c acting in conjunction with the grid-cathode capacitance. The inductance is partly due to the cathode structure itself but arises principally in the lead to earth via the valve pin. The inductance may be reduced by the use of two or more cathode leads simply connected in parallel. Much better results are obtained (a 50% increase in R_g) if the input and output circuits, though nominally having a common earth, are in fact separated, one cathode lead being assigned to each. The coupling between cathode current and grid voltage is then much reduced, since the common impedance is lower.*

(b) The effect, discussed for example by FERRIS [502], of the finite transit-time of an electron between leaving the cathode and passing through the grid is to lower the input resistance. The influence on the input capacitance is small.

The transit-time may be reduced by constructing the valve with a grid of many very fine wires situated extremely close to the cathode. The close spacing also gives a higher value of mutual conductance, which is very desirable, but leads to an increase in grid-cathode capacitance. The improvement obtained by virtue of the increase in R_g however more than outweighs the latter disadvantage.

---

* This illustrates the general care that must be taken with earth connections in the practical construction of any circuit involving high frequencies. It is important to realize that the surface of a metal chassis possesses some kind of impedance and currents flowing over it will therefore produce potential differences which may appear in adjacent parts of the circuit with adverse effects on performance. It is often best to adopt the technique, essential in amplifiers of even moderately high gain, of collecting together at a single point (insulated from the chassis) all the earth connections associated with a particular valve. The now limited number of earth points may then be connected together by a heavy insulated copper strip which is joined to the chassis as close as possible to the input plug.

5.2.1.4 Output conductance

A small effect occurs in the anode circuit due to the inductance of the anode, screening and suppressor grids and their connections. Phase shifts occur resulting in the appearance of a small shunt conductance which is negligible in practice.

5.2.1.5 Anode-grid capacitance

In a pentode valve the capacitance between anode and grid is very small. Care must be taken however to avoid capacitance in the wiring between these two points. The familiar Miller effect usually gives rise to negative voltage feedback with a consequent frequency dependent reduction in gain. Under certain conditions, when appreciable stray inductance is present, the feed-back may become positive and instability will result.

At high frequencies, the capacitance is apparently modified by the interelectrode capacitances of the anode, suppressor, screening and control-grids, associated with their stray inductances. The effective capacitance may even become negative, again resulting in instability.

The importance of proper de-coupling of the screening-grid may be pointed out here. The capacitance between grid and screen is many times that between grid and anode and the Miller effect may occur at the screen with the undesirable results just mentioned. Non-inductive capacitors* must be employed for decoupling, and everywhere else for that matter, and low value carbon resistors may be inserted freely (except perhaps when the added stray capacitance to earth would be disadvantageous) to damp any resonances between capacitance and stray inductance. Ceramic capacitors are usually available of value large enough for screen decoupling purposes; a resistor value of 100 Ω is suitable. Ceramic capacitors of sufficiently high value for use in decoupling anode and H.T. supply lines are not available, however, and an electrolytic type may be connected in parallel, provided a 10 Ω carbon resistor is inserted in series with the latter.

5.2.2 Valve Requirements—Figure of Merit.

In addition to minimizing the effects listed above, valve design aims primarily at obtaining a high value of mutual conductance, to

---

* High permittivity ceramic capacitors, of the small button type, are satisfactory, but MOODY et al. [503] point out that care must be taken to ensure that mechanical resonance of the ceramic, excited by piezo-electric effects, does not occur within the band of signal frequencies (DEIGHTON [504]).

which the stage gain is proportional, when use as a voltage amplifier is envisaged. The mutual conductance increases with the standing anode current and therefore with the screen-cathode voltage; high power dissipation, particularly at the screen, should therefore be admissible. In some applications there is a requirement for low noise.

When a valve is to be used in trigger circuits further requirements must be met. There is a demand for a high peak cathode emission current, to be obtainable with as low a grid drive current as possible; a short grid base, i.e. a sharp grid cut-off characteristic, is also advantageous.

As one might expect, when it comes to valve design, some of these requirements are found to be mutually conflicting. Broadly speaking, a large value of the mutual conductance implies high operating voltages and currents and therefore large electrodes, to handle the dissipation involved, with a consequent increase in electrode capacitance. Several figures of merit are in use for assessing the suitability of a valve for a particular purpose. For our present application the wide-band figure of merit F is appropriate; this quantity is defined by:

$$F = \frac{g_m}{2\pi(C_a + C_g)} \qquad (5.1)$$

When g_m is in mA/V and the capacitances in pF then F gives the frequency in KMc/s at which the valve might produce a voltage gain of unity when driving its own total electrode capacitances.*

In the case of a resistance coupled stage, of gain $g_m R$, feeding an identical stage, the amplification falls at the frequency $1/2\pi(C_a + C_g)R$ to $1/\sqrt{2}$ times its normal value at medium frequencies (3 db loss). The product of gain and bandwidth from zero up to this frequency is accordingly equal to the figure of merit.

The following simple analysis illustrates the significance of the figure of merit in pulse amplifiers. Suppose we want to obtain an overall gain G, with valves of a given F, so that the output pulse rise-time (for a step function input) is to have the value t_r. We wish to know the number N of valves required when identical resistance-capacitance coupled stages are connected in cascade.

---

* WHEELER's figure of merit uses $\sqrt{C_a . C_g}$ instead of the sum; the sum is, however, the relevant quantity when the valves are connected in cascade and have resistive anode loads. Other writers omit the factor 2π; the figure then has less physical significance than as defined in the text.

AMPLIFIERS

If R is the anode load of each stage then

$$G = (g_m R)^N \tag{5.2}$$

When $C = C_a + C_g$ is the total capacitance the rise-time of one stage, for a step function input, may be taken as CR. The resultant rise-time for the whole amplifier is therefore (see equation 1.6)

$$t_r = CR\sqrt{N} \tag{5.3}$$

For the figure of merit we have:

$$F = \frac{g_m}{2\pi C} \tag{5.4}$$

It follows at once that

$$G^{\frac{1}{N}} \sqrt{N} = 2\pi F t_r$$

or

$$\frac{1}{N} \cdot \log G + \frac{1}{2} \cdot \log N = \log 2\pi F t_r \tag{5.5}$$

N is thus determined when the quantities G and Ft_r are given. When g_m (or C) is known then R may be found from relations 5.2 (or 5.3).

The figure of merit, useful criterion as it is, must be used with reserve. If, for example, a valve has $F = 100$ Mc/s, a gain of 20 will be obtained up to a frequency of 5 Mc/s. It does not follow however that the valve could therefore be used to give a gain of 2 up to 50 Mc/s, since transit-time effects and lead inductance might become relevant at the higher frequency. Again, the figure of merit obtained from manufacturers' valve data does not allow for stray capacitance in the wiring. If there is a choice between two tubes with the same figure of merit, the one with larger capacitances should be selected since the effect of wiring strays will then be proportionately less.

It is worth pointing out that, when a number of identical valves are connected with all their electrodes in parallel, the wide-band figure of merit of the combination is the same as that of one valve. There is thus no advantage to be gained by connecting parallel groups of valves in cascade. Parallel connection of valves is only of use, for example, in an output stage which has to feed into a given low impedance. The increase in current handling capacity will then enable a larger output to be obtained without additional deterioration in rise-time (which will be principally determined by the previous stages).

When the gain per stage is low, and the number of valves consequently large, the H.T. supply power required may be inconveniently high (this applies particularly to the distributed amplifiers described in § 5.4). An economy figure of merit defined by the ratio g_m/I_a, where I_a is the standing anode current, is relevant in such cases. This criterion may well determine the choice between two valves of similar wide-band figures of merit.

5.2.3 Some Valve Types.

Full details of secondary emission types of valve may be found in references [429]–[433] and the salient properties of these, and of some other valves of interest to us, are given in Appendix III.

FORD [505] points out that up to the year 1949 the familiar types 6AK5 and 6AC7 were the most suitable for use in wide-band voltage amplifiers, and an improved pentode tube, the Western Electric type 404A, is then described in his paper. A further article by FORD and WALSH [506] gives information on various tubes, including the 404A, and on the 408A, which differs from the 6AK5 solely in respect of heater voltage and current. The type of construction alluded to in § 5.2.4(b) is fully detailed in these papers and in an earlier one by WALSH [507].

WALLMARK [508] describes a promising new experimental tube in which current density control and beam deflection control are simultaneously exercised by an input grid. Signals on the control grid cause variations in space charge density, in the normal manner, and it is arranged that the variation in charge density in the emerging beam disturbs a transverse electric field, thus altering the beam direction. Part of the beam is stopped by an intercepting electrode and the remaining portion, which varies both in density and in cross-section, travels onwards to a dynode electrode. A high degree of control is therefore obtained and a large output is available from the final anode which collects the secondary electrons emitted from the dynode. Some data on this tube are given in Appendix III.

5.3 INTERSTAGE COUPLING IN CASCADE AMPLIFIERS

We now turn to a consideration of one or two types of interstage coupling circuit which are designed to give uniform amplification over a larger frequency band than does simple resistance-capacitance coupling. A general theoretical account of the problem is given by

HANSEN [509], and particular circuits are discussed by HEROLD [510] and by SEAL [511]; a paper by BEDFORD and FREDENDALL [512], for example, shows how the calculation of the response of the system may proceed in certain practical cases.

Coupling networks are designed primarily to compensate for the effects of valve anode and grid capacitances, plus wiring capacitance. The simplest modification which can be applied to the circuit of fig. 5.1 is to insert an inductor in series with the anode load resistor such that the resonant circuit so formed is approximately critically damped.

5.3.1 Practical Circuit.

A more refined circuit, used by MOODY and his collaborators [503], is shown in fig. 5.2(a) and is by far the best of all the simple coupling

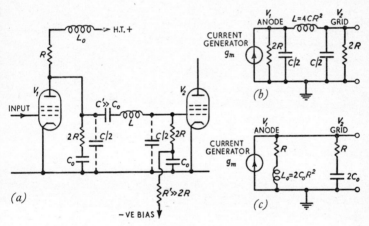

Fig. 5.2. (a) Intervalve filter type coupling.
(b) Equivalent circuit at low frequencies.
(c) High frequency equivalent.

networks. We have seen in § 2.4 how a low-pass filter section possesses a resistive characteristic impedance and an inductor L is accordingly connected to form such a section in conjunction with the stray and interelectrode capacitances $C/2$ (shown dotted). The inductance L_0 and the capacitance C_0 are large, and consequently the circuit is equivalent to that depicted in fig. 5.2(b) at medium and high frequencies. If $2R = \sqrt{L/C}$ the section is properly matched at both ends—a condition which we would naturally expect to be desirable* (though since only one section is involved the question of reflected "waves" hardly arises). The effective anode load is thus equal to $2R$ in parallel with the characteristic impedance $2R$ of the filter section, and, since the valve behaves as a generator of constant current (when R is very much less than the anode slope resistance), the gain to the grid of the next stage V_2 is equal to $g_m R$.

A blocking capacitor C' must be included between the anode of V_1 and the grid of V_2 in order that the normal grid bias may be applied, via the resistor R', to the second valve. The "dead" ends of the resistors $2R$ must be A.C. connected to earth, through capacitors C_0. Now it is difficult, in practice, to obtain a capacitor of large value, to pass the low frequency components, and yet possessing a very low stray inductance as is required for the impedance to be low at high frequencies. The maximum value of the capacitance of the bypass condensers C_0 is limited and a rise in gain would therefore occur at low frequencies. The inductance L_0, however, causes the gain to fall off at low frequencies and the two effects can be made to cancel. Fig. 5.2(c) shows the equivalent circuit at low frequencies where it is assumed that $C' \gg C_0$.† When the component values are related as shown the network presents a constant impedance R, which is purely resistive, to the anode of the valve V_1; the gain therefore remains at the value $g_m R$‡ right down to the very low frequency at which $wC'R'$ ceases to be much greater than unity.

The high frequency limit to the stage may be determined by the cut-off frequency of the filter section or by the input conductance of the next valve. Since the anode capacitance of a valve is usually less than the grid capacitance, it should be arranged that the wiring capacitance lies as far as possible on the anode side of the inductor L which should be mounted as close as possible to the grid of V_2. For the same reason the blocking capacitor C' should be inserted on the

---

* Perhaps one of the optimum resistance terminations worked out by GIACOLETTO [513] might be employed with advantage—see § 2.4.1.

† A considerably larger value for C' is admissible since the capacitor is connected in series with a coil and the presence of stray inductance is accordingly immaterial.

‡ A factor of two improvement in gain would be obtained over the whole band, though with some deterioration in high-frequency performance, if the "matching" resistor $2R$ and the capacitor C_0 at the value anode were omitted. Constant impedance would be maintained down to low frequencies provided the anode load resistor, which is connected in series with the inductance L_0, were raised to the value $2R$.

anode side, so that the effect of its stray capacitance to earth may be felt here rather than on the following grid.

MOODY and his co-workers have constructed a four stage amplifier employing EFP60 type valves, with coupling component values of $R = 160\ \Omega$, $L = 1 \cdot 8\ \mu\text{H}$, $L_0 = 50\ \mu\text{H}$, and $C_0 = 1000$ pF. An overall voltage gain of 500 was obtained, corresponding to a gain per stage of 4.7, over a frequency band extending from 10 Kc/s to nearly 50 Mc/s; the gain-bandwidth product per stage is accordingly approximately equal to 240 Mc/s. The gain-bandwidth product of a single resistance-capacitance coupled stage is given by the figure of merit $F = 225$ Mc/s (see Appendix III). This value is based on a consideration of the valve interelectrode capacitances alone and is therefore optimistic; the value found above for the practical amplifier employing filter sections, which of course includes the effect of wiring strays, shows the superiority of the latter type of coupling.

5.3.2 Use of Dynode in Secondary Emission Valves.

As already mentioned in § 4.3 secondary emission valves offer a unique facility, not possessed by conventional tubes, whereby

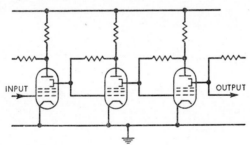

Fig. 5.3. Dynode coupled amplifier (blocking capacitors not shown).

amplification may be obtained without phase inversion if the output is taken from the dynode instead of from the anode. In many nuclear physics applications, for example, we require to amplify pulses in a certain amplitude range in the presence of much larger pulses. In the later stages of a conventional anode to grid coupled amplifier paralysis ensues owing to the passage of grid current in alternate valves; this effect is most undesirable since the amplifier remains insensitive for a comparatively long and indeterminate period. No paralysis occurs with the circuit indicated in fig. 5.3

provided the input pulses are negative going. When a large input signal is applied the later stages simply are cut-off for the duration of the pulse and the amplifier returns to normal almost immediately after the pulse is over. Such an amplifier has a smaller gain-bandwidth product, owing to the lower mutual conductance to the dynode and the somewhat greater capacitance of this electrode. This system should therefore only be employed in the later stages of a high gain amplifier where overloading would be likely to occur.

5.3.3 General

Some remarks of general application to circuits employing secondary emission valves may be interpolated here. Such tubes are

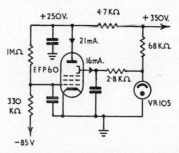

Fig. 5.4. D.C. supply network for EFP60.

unstable in operation, due to variations in the secondary emission coefficient of the dynode, and Moody points out that proper operating conditions are not necessarily set up simply by applying the nominally correct potentials to the dead ends of the anode and dynode load resistors. It is important to include some form of D.C. negative feed-back in each stage separately in order that the anode current becomes self-determining at the desired value and to assist in stabilizing the mutual conductance to which the stage gain is proportional. The feed-back is applied between anode and grid, as shown in fig. 5.4, and provision is also made for stabilizing the dynode operating conditions (this electrode requires to be maintained at a potential of about 150 V).

The circuit of fig. 5.4 is only intended to operate at D.C or very low frequencies and is subsidiary to, and must not interfere with, the normal functioning as a high-speed amplifier. On the other hand, as depicted in fig. 5.3, the feed-back from anode to grid may be allowed to act at all frequencies, thus providing improved gain stability or improved bandwidth.

BORG [514] has constructed an amplifier employing a modified form of low-pass Π-filter section for the interstage coupling. A gain of 1000 is obtained over a frequency band extending from 11 to 118 Mc/s using nine secondary emission valves type VX5038* in cascade. When the same number of 6AK5 tubes was employed to give a similar gain a bandwidth extending from 7 to 78 Mc/s was obtained.

A. B. GILLESPIE, at A.E.R.E. Harwell, has recently designed a cascade amplifier using VX5038 valves arranged in pairs. Negative feed-back is applied from the anode of the second valve in each pair to the cathode of the first. A voltage gain of 30 per pair is obtained up to a frequency of 30 Mc/s; alternative component values enable a gain of 6 to be achieved up to 60 Mc/s.

5.4 DISTRIBUTED AMPLIFIERS

We have seen how the gain-bandwidth product of a resistance coupled amplifier stage is a constant which depends only on the type of valve employed (if cathode lead inductance and certain other effects are neglected). By tolerating a smaller gain per stage at medium frequencies, the bandwidth can be increased and a very wide frequency band handled if the gain were allowed to fall below unity. In such an event it is clearly futile to connect stages in cascade since the gains of the separate stages are multiplicative; we accordingly seek to arrange that the gain contributions of each stage become additive. As pointed out earlier, the simple parallelling of valves produces no improvement in overall gain-bandwidth product but it is possible to parallel the valves in a special way such that the mutual conductances effectively add but the self capacitances do not. The basic principle is to allow the grid and anode capacitances to form the shunt capacitative elements of two lumped delay lines as shown in fig. 5.5.

A signal applied at the input to the grid line travels along it and passes the grid of each valve in turn. Anode current flows in turn and the pulse at each anode divides into two wave components, one of

---

* This valve has a higher wide-band figure of merit than the EFP60 and is therefore more suitable for use in linear voltage amplifiers. The latter valve is preferable in trigger circuit applications, however, when a very high pulse current is required.

which progresses forwards towards the output, and the other backwards. If the delay times per section in the two lines are the same the anode signals travelling forwards all add together without relative delay; those moving in the backward direction are absorbed by a matching resistor. The end of the grid line at the output end of the amplifier is matched. If Z_{0a} is the characteristic impedance of the anode line (as viewed looking in at the capacitative elements) the actual load resistance seen by each valve is equal to $Z_{0a}/2$; the

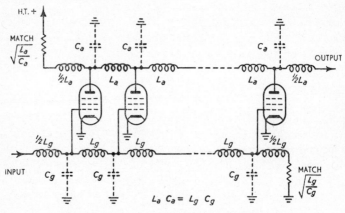

Fig. 5.5. Principle of distributed amplification.

gain produced by n valves, each of mutual conductance g_m, accordingly has the value $ng_m Z_{0a}/2$.

The idea of effectively separating the electrode capacitances, by means of inductances, whilst allowing the mutual conductances of the individual tubes to add together is originally due to PERCIVAL [515]; ESPLEY [516] has used an amplifier operating on a similar principle. The first complete account, however, of the theory and practice of such an arrangement is due to GINZTON* et al. [517] and HORTON et al. [518] (also see STEINBERG [519]).

Uniform amplification can be obtained down to audio frequencies and even to D.C. if desired. The high frequency limit may be determined by either (i) the cut-off frequencies of the anode and grid lines (which are identical, since the delay times per section in the two lines must be the same) or (ii) effects within the valves themselves, other than the simple grid-earth and anode-earth capacitances. A

---

* The term "distributed amplification" has come into common useage but it should be remembered that all the circuit elements are lumped.

simple account will now be given for the case when the former alternative is the limitation, and effects of the second kind will be discussed in a later section.

5.4.1 First Order Theory.

Before we can arrive at an optimum design we must first decide on how the input and output impedances are to be related. An obvious choice is to make them equal to facilitate the connecting of a number of stages in cascade (each stage will comprise several valves arranged in distributed fashion). Again, if a significant gain-bandwidth product is to be arrived at comparisons of gain must be made at the same impedance level since a pure transformer voltage amplifying effect should be excluded from such a criterion of performance. Let us therefore first consider the case where the anode line is simply terminated by a load resistance which is equal to the input (characteristic) impedance of the grid line but is in general different from the characteristic impedance of the anode line.

If suffixes g and a refer to the grid and anode lines respectively the transmission coefficient at the output end of the anode line (see § 2.2.4) is $2Z_{0g}/(Z_{0g} + Z_{0a})$. The overall gain G is therefore given by

$$G = \frac{2Z_{0g}}{Z_{0g} + Z_{0a}} \cdot \frac{ng_m Z_{0a}}{2} \qquad (5.6)$$

Now the characteristic impedances and cut-off frequency given in § 2.4.1 are

$$Z_{0g} = \sqrt{\frac{L_g}{C_g}}, \qquad Z_{0a} = \sqrt{\frac{L_a}{C_a}} \qquad (5.7)$$

$$f_c = \frac{1}{\pi\sqrt{L_a C_a}} = \frac{1}{\pi\sqrt{L_g C_g}} \qquad (5.8)$$

and, if we assume that no extra capacitors are added so that C_g and C_a are the capacitances of the valves (including wiring strays), the figure of merit, equation 5.1, may be introduced. It is then easily shown that

$$G = \frac{2nF}{f_c} \qquad (5.9)$$

Thus, if f_c and G are predetermined, nF is found. Clearly a type of valve with a high wide-band figure of merit should be selected the choice being influenced by a consideration of the economy figure of merit if a large number of valves are going to be required.

In a practical case the value of f_c is given and suitable valves are selected, thus determining C_g and C_a. The impedances and inductances of the lines are then determined by the relations

$$Z_{0g} = \frac{1}{\pi f_c C_g}, \qquad L_g = \frac{1}{\pi^2 f_c^2 C_g} \qquad (5.10)$$

$$Z_{0a} = \frac{1}{\pi f_c C_a}, \qquad L_a = \frac{1}{\pi^2 f_c^2 C_a} \qquad (5.11)$$

If the values of Z_{0g} and Z_{0a} turn out to be inconveniently low, valves with smaller interelectrode capacitances must be chosen.

5.4.1.1. *Optimum Grouping of Valves*

It is noted that there is no optimum choice for the value of the characteristic impedance of the anode line; we have simply to choose a valve with a high figure of merit and the rest follows. There is, however, an optimum way in which a total of N valves may be grouped, in m identical cascaded stages, each containing n valves, such that N is a minimum for a given overall gain A, say. If G is the gain of each stage then

$$A = G^n = \left(\frac{2nF}{f_c}\right)^m \qquad (5.12)$$

Solving for n we have

$$n = \frac{f_c}{2F} A^{1/m}$$

Hence

$$N = mn = \frac{m f_c}{2F} A^{1/m}$$

Differentiating this with respect to m and putting $dN/dm = 0$ it is found that N is a minimum when $m = \ln A$. It then follows that

$$G = e = 2 \cdot 72 \quad \text{and} \quad n = 1 \cdot 36 f_c/F \qquad (5.13)$$

5.4.1.2 *Alternative forms of Interstage Coupling*

The cascade arrangement suggested above, in which the output from each anode line is fed directly into the next grid line (or the final load) the impedance of which is in general different from the characteristic impedance of the anode line, is open to the objection that reflection takes place at the mismatch. If the reverse end of the anode line is perfectly matched then all is well, but in practice this will not be the case, and there is a danger of small secondary pulses appearing at the output after a time delay equal to the double transit-time on the line. In applications where we are primarily interested in the leading edge of a fast pulse, such reflections will arrive too late to have any adverse effect, and the simplicity of this comparatively crude method of coupling has much to recommend it.*

An improvement of a factor two in gain, over the matched case, can be realized if a single buffer valve is inserted between the output end of the anode line and the final load. The input capacitance of this valve will, in general, be greater than the anode capacitance of any one of the preceding valves, and the extra filter section will therefore require a higher inductance, in order to keep the characteristic impedance of this section equal to that of the anode line. This section thus has a lower cut-off frequency than the main line but the additional distortion can usually be tolerated, since there is only one such section. The use of a buffer valve is only advantageous if the desired bandwidth is not too large, so that it can, in fact, be realized at the buffer output, without the necessity of employing an anode load resistor so low in value that the gain of this valve falls below unity.

If one is apprehensive, particularly at frequencies above 100 Mc/s, about reflections from the mismatch at the output an obvious alternative is to arrange that the impedances of the anode and grid lines are equal. Since the anode capacitance of an ordinary valve is less than the grid capacitance additional capacitors will need to be added between each anode and earth. The nominal cut-off frequency of both lines is the same but we know that phase distortion sets in before the cut-off frequency is reached. A policy of deliberately increasing the anode capacitances therefore seems to be in the wrong direction and a preferable arrangement is indicated in fig. 5.6. One extra filter T-section is added between each valve in the anode line, composed of capacitances C_a' and inductances L_a'. Since the impedances of all sections are to be the same we must have

$$\frac{L_g}{C_g} = \frac{L_a}{C_a} = \frac{L_a'}{C_a'} \tag{5.14}$$

---

* GINZTON et al. go to the other extreme and consider the case in which the anode line of one stage is matched to the grid line of the next by means of a transformer. Transformers have been used with success but in most applications one would prefer to avoid the further complication introduced even at the possible expense of having to add a few extra valves.

and, since the time delay between grids is to equal that between anodes, the following relation must also be satisfied

$$\sqrt{L_g C_g} = \sqrt{L_a C_a} + \sqrt{L_a' C_a'} \qquad (5.15)$$

The quantities C_g and C_a are determined by the valves used and a knowledge of the desired cut-off frequency in the grid line then

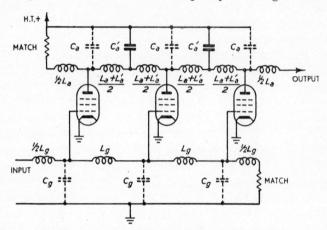

Fig. 5.6. Arrangement for making anode and grid line impedances equal.

determines L_g (relation 5.10). The quantities L_a, L_a' and C_a' are then determined as follows from equations 5.14 and 5.15:

$$L_a = \frac{C_a}{C_g} L_g \qquad (5.16)$$

$$L_a' = L_g \left(1 - \frac{C_a}{C_g}\right) \qquad (5.17)$$

$$C_a' = C_g \left(1 - \frac{C_a}{C_g}\right) \qquad (5.18)$$

The cut-off frequencies of the two kinds of section comprising the anode line are each greater than that of the grid line and, although the total number of anode sections is doubled, there is a resultant reduction in phase distortion.

Another possibility presents itself. It might be thought desirable to include a single resistor of value equal to $(Z_{0a} - Z_{0g})$, on the assumption that $Z_{0a} > Z_{0g}$, in series with the junction between the anode line and the following grid line or output load. The system would then be matched in the outgoing direction (see § 3.2). An attenuation of Z_{0g}/Z_{0a} will occur at the junction between the lines and the gain from the grid line of one stage to the grid line of the next will be equal to $\dfrac{ngZ_{0a}}{2} \cdot \dfrac{Z_{0g}}{Z_{0a}}$. This result is just the same, however, as that which obtains when the characteristic impedances of the two lines are made equal in the first place.

A further arrangement used originally by ESPLEY [516] and discussed by GINZTON et al. [517] may be employed in cases when the output impedance required is considerably lower then the anode line impedance. The anode delay line sections are tapered in harmonic progression such that if Z is the impedance of the anode section associated with the first valve the impedance of the next section is made equal to $Z/2$ and the next $Z/3$ and so on. The part of the forward going signal reflected at each change in impedance is exactly cancelled by the backward travelling signal component generated at the anode of the valve in question and there is no resultant wave in the reverse direction. The terminating resistor at the reverse end of the line may be omitted and all the power reaches the load. If there are n valves in the stage, the impedance of the last section is made equal to Z/n and the load should have the same value.

5.4.1.3 Output Stage Design

In order that the output signal should by directly proportional to the input voltage all the valves must operate on the linear part of the characteristic curve of anode current versus grid voltage. The signal amplitude in the grid line of the last stage of a composite amplifier must therefore not exceed a few volts. The optimum cascade arrangement discussed above, in which the gain of each stage is arranged to be equal to 2.72, may be employed in the earlier parts of the complete amplifier; when a large output is required, however, there is no alternative but to incorporate the necessary number of valves in the last distributed stage. The choice of valve type for the output stage should be governed by a consideration of the total anode current change, over the linear region, which the valve is capable of delivering, rather than by the mutual conductance alone. Thus a high slope valve which is linear only over a small change in grid potential would be suitable for use in the early stages of a composite amplifier, but one with a somewhat lower slope and a much wider acceptable grid variation would be preferable for the output stage.

If the output stage is to accept negative going input signals, a positive output being obtained, the grid bias must be adjusted as near to cathode potential as considerations of dissipation within the valves will allow. A large amount of H.T. power* is consumed and the peaks of large signals will become levelled off due to the non-linear approach to cut-off. If, on the other hand, the input signals are positive the valves may be biassed nearly to cut-off and several advantages are gained:

(a) a saving in H.T. power is obtained or, alternatively, the screen potentials may be raised, without exceeding the valve rating, thus increasing the grid swing over which the anode current varies linearly;

(b) the valve characteristic is only non-linear for small pulses and larger signals suffer no distortion over most of their excursion;

(c) the valves may even be driven into a small amount of grid current, if the characteristic impedance of the grid line is low enough; greater output again results.

Alternate stages in a composite amplifier must accept pulses of different polarity and the grid bias on each stage should be set to different values accordingly, but it should always be arranged, if at all possible, that the output stage accepts positive signals. The transmission-line type of pulse inverter described in § 3.5 is particularly suitable for use at the amplifier output, when one is primarily interested in measuring peak amplitude and when the measuring equipment requires a positive going input pulse.

An interesting alternative suggested by MOODY and his collaborators [503] may be adopted when output pulses in opposite phase to one another are required simultaneously. Secondary emission valves are employed and the control grids, anodes and dynodes are connected to three separate delay lines (fig. 5.7). The delay-times per section of all three lines are made the same, and by suitably adjusting the impedances of the anode and dynode lines a push-pull output may be obtained. It should be arranged, of course,

---

* The pedestal pulse technique mentioned in § 4.2.2.2 may be employed here.

that a positive going signal is applied to the grid line. Such an output stage would be useful for providing symmetrical deflection to a cathode ray tube, for example, and the number of valves required

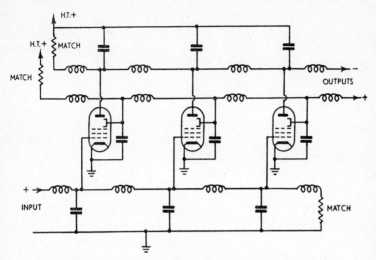

Fig. 5.7. Push-pull output stage using secondary emission valves.

is less than if two separate amplifiers, using conventional valves, were employed.

5.4.2 Further Considerations.

5.4.2.1 Derived Filter Networks

As was pointed out in § 2.4.2, certain modified forms of low-pass filter have a delay time per section which is less dependent on frequency than is the case with the simple constant-k type envisaged above. In practice m-derived filters are usually employed and optimum delay characteristic is obtained when $m = 1 \cdot 27$, corresponding to a coefficient of coupling between half-coils of $0 \cdot 23$ (negative). The values of the characteristic impedance and the inductance L (see fig. 2.17(d)), for a given capacitance C and cut-off frequency f_c, are obtained from relations 2.72 which may be put into the form

$$Z_0 = \frac{0 \cdot 39}{Cf_c}, \quad L = \frac{0 \cdot 13}{Cf_c^2} \quad (5.19)$$

An improved match* is obtained when an m-derived half-section, with $m = 0{\cdot}6$, is inserted before a pure resistive termination as depicted in fig. 2.18(b). The quantities L_k and C_k are defined in terms of the actual L and C as shown in fig. 2.16(d); the value of m here refers to the main line.

The bridged T-network of fig. 2.19(a) has been successfully employed and has the advantage that the matching impedance required is purely resistive. The coils should again be wound in the same direction and the coefficient of coupling may have a value of about $0{\cdot}2$. This network has the disadvantage that an extra component, namely the bridging capacitor, has to be added to each section; the value also requires experimental adjustment to obtain optimum practical performance.

5.4.2.2 *Low Frequency Limitations*

When working at the high frequencies with which we are concerned it is desirable to arrange that the cathodes of all valves are at earth potential, since non-inductive high capacitance condensers, which might be used to by-pass the cathode resistors, cannot be obtained in practice. The grid line must therefore be insulated from earth, as far as D.C. is concerned, in order that grid bias may be applied. Blocking capacitors C_1, C_2 and C_0, shown in fig. 5.8(a), are therefore required. The capacitors C_0 must be non-inductive and their value is accordingly limited. At low frequencies their reactance becomes of the order of the characteristic impedance Z_{0g} of the grid line and the amplitude of the grid signal is accordingly greater than that at high frequencies. At the lower frequencies the delay time of the lines may be neglected and the circuit reduces to that shown in fig. 5.8(b), where the three valves have been drawn as one. If the internal resistance of the source of e.m.f. E feeding the grid line of the first stage is equal to Z_{0g}, then it is easily seen that the signal in the line remains at the value $E/2$, independent of frequency, provided we make $C_1 = C_0$. The signal will fall off

---

* WELLS has drawn attention to the fact that in applications where one is primarily concerned with the leading edge of a pulse it is better not to attempt a match in the grid line immediately after the last valve in the stage. A number of extra sections may be added, embodying capacitors to simulate the presence of valve grid capacitance, before the termination is reached. The reflected pulse from an imperfect termination is then delayed before it again reaches the grid of the last valve and travels in the reverse direction down the grid line. The frequency response of such an amplifier would appear to be poor, but a clean leading edge is in fact obtained.

however at very low frequencies when $2\omega C_0 R'$ ceases to be much greater than unity.

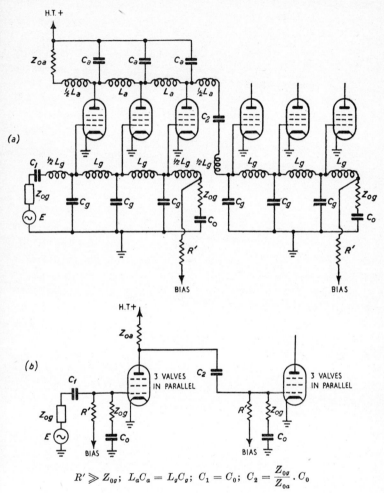

Fig. 5.8. Compensation for low-frequency distortion. Circuit (b) is equivalent to (a) at low frequencies.

Similarly, at low frequencies, the anode line, when correctly terminated in the reverse direction by a resistor Z_{0a}, behaves as a source of e.m.f. equal to $ng_m V_g Z_{0a}$ and internal resistance Z_{0a} (V_g is the grid signal). It is easily shown that the signal in the grid line of the second stage is equal to $ng_m V_g Z_{0a} Z_{0g}/(Z_{0a} + Z_{0g})$ provided $C_2 = \dfrac{Z_{0g}}{Z_{0a}} C_0$. This expression for the gain is the same as that applying at high frequencies.

A practical difficulty sometimes arises in connection with the anode line reverse termination. If the full anode current is allowed to flow through this resistor, the high dissipation may call for the use of a wire-wound resistor of large physical size; the accompanying

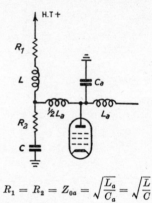

$$R_1 = R_2 = Z_{0a} = \sqrt{\dfrac{L_a}{C_a}} = \sqrt{\dfrac{L}{C}}$$

Fig. 5.9. Compensation of inductance in resistive termination.

large stray capacitance to earth is undesirable and such resistors are also inductive. A high-frequency choke may be connected in parallel with the termination, to pass the steady anode current, thus allowing a small carbon resistor to be employed. Such an arrangement has been used by MYERS [525] in which the choke was actually wound on top of a carbon film type resistor. Clearly this choke will give rise to a reduction in gain at the lower frequencies. Several chokes may be connected in series, graded to give increasing inductance towards the H.T. supply end of the resistor. The increased self capacitance of the larger chokes is immaterial, since smaller chokes are connected between them and the signal side of the termination.

The network shown in fig. 5.9 however, should give superior performance with few extra components. The resistors R_1 and R_2 are both equal to the characteristic impedance Z_{0a} of the anode line and, if L and C are chosen such that $Z_{0a} = \sqrt{L/C}$, the network presents a constant resistance equal to Z_{0a} down to zero frequency.

At high frequencies, R_2 is the effective termination and this may be a small carbon resistor; the condenser C is of comparatively low value and must be non-inductive. The resistor R, which passes the total anode current, may be large and of the wire-wound type; its inductance does not matter since an inductance is already connected in series with it and its large stray capacitance to earth is also irrelevant since, at high frequencies, the inductance L effectively isolates the signal line from this capacitance.

5.4.2.3 Limitations at High Frequencies

We have so far assumed that the useful upper frequency limit of the amplifier is determined by the cut-off frequency of the delay lines. This may indeed be the case but at frequencies in excess of 100 Mc/s the effect of grid loading, connecting lead inductance, and line loss, must be taken into account. Such effects are discussed by GINZTON et al. [517] and in further detail by HORTON et al. [518] and it appears that grid loading is the dominating limitation.

Inductance in the grid and anode leads can be compensated by suitably adjusting the m-derived filter sections but cathode lead inductance L_c is more serious. It can be shown that this inductance, acting together with the grid-cathode capacitance C_{gc}, gives rise to a shunt conductance G per section of grid line. The effect of the electron transit-time τ is to increase the conductance which varies with frequency according to the relation

$$G = g_m \omega^2 (L_c C_{gc} + K\tau^2) \qquad (5.20)$$

where K is a constant depending on the particular type of valve and g_m is the mutual conductance. The input resistance of a 6AK5 valve, for example, has a value of the order of 250 Ω at a frequency of 400 Mc/s. It has been shown that the fractional loss in the gain per stage is approximately equal to

$$1 - \frac{AG}{2g_m} \qquad (5.21)$$

where A is the stage gain, neglecting losses. The number of valves enters into this expression by virtue of the ratio A/g_m and it is clear that there is no advantage to be gained by increasing the number beyond a certain limit, owing to attenuation in the grid line.

The design of amplifiers for use up to frequencies of the order of 400 Mc/s has also been discussed by WEBER [520]. Before considering methods of compensating for the valve input conductance it is useful to see how the effect of cathode lead inductance may be reduced. It has already been indicated in § 5.2.1.3 that a valve with more than one cathode lead may be employed with advantage, so that the input and output currents do not flow through a common lead inductance. The 6AK5 type of valve has two such connections and the arrangement indicated in fig. 5.10 has been employed with success by WEBER. One cathode lead from each valve is earthed directly to the chassis. The other cathode leads are connected to a straight piece of wire, virtually a third transmission line, which is arranged to have a much greater capacitance to the anode line than

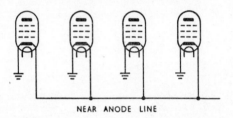

Fig. 5.10. Arrangement of connections for reducing the effect of cathode lead inductance (after WEBER).

to the grid line. By this means the majority of the output current from the first valves is isolated from the input current. This improved method of wiring has no influence, of course, on the contribution to the input conductance due to transit time.

A simple method of neutralizing the input conductance, when the valve possesses only one cathode connection, is to connect a small resistor in parallel with a small capacitor between the cathode of the valve and earth. This combination appears in series with the cathode lead inductance and the whole will present a nett capacitative reactance, up to a certain frequency, and no positive input conductance will be obtained. This circuit has been employed by W. R. HEWLETT but the gain of the amplifier is somewhat reduced over the working frequency band.

Neutralization may also be achieved, without appreciable reduction in gain, by connecting a small inductance in series with the screen connection. The cathode, control-grid and screening-grid behave as a triode valve with an inductive load and, by virtue of the grid-screen capacitance, a negative input conductance results which tends to cancel the positive input conductance, due to cathode lead inductance and transit-time. The value of the screen inductance must be chosen such that no resonance with the screen capacitance occurs within the working band.

5.4.2.4 Gain Stability

An advantage possessed by the distributed amplifier is the *comparative* freedom from any tendency to self-oscillation. The

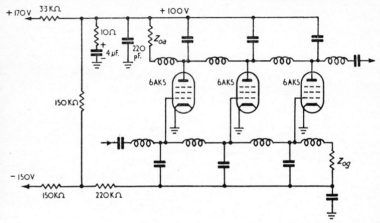

Fig. 5.11. Stabilization of gain by D.C. stabilization of standing anode current.

arrangement is also reliable, since the failure of a valve merely results in a slight loss in gain. A major disadvantage in work involving the precision measurement of pulse amplitudes is the variation in gain which occurs due to changes in mutual conductance. In microsecond amplifiers overall negative feed-back is usually employed to improve linearity and stabilize the gain but this cannot be done in the case of distributed amplifiers.

Improved stability results if the screen potentials and heater supplies are stabilized. In addition (or instead), steps may be taken to arrange that the total anode current, upon which the mutual conductance principally depends, is maintained constant. Some form of negative feed-back, operating only at zero or very low frequencies, is called for and one arrangement is indicated in fig. 5.11. In a circuit of this sort it is necessary that all the valves in the stage be approximately matched as regards their D.C. characteristics.

5.4.2.5 Noise

A distributed amplifier suffers from the usual sources of noise* to which wide-band amplifiers are prone. Contributions to the total noise output arise from:

(a) thermal noise in the output impedance of the source feeding the grid line, and in the grid and reverse anode terminating resistors,

(b) shot and partition noise in the valves,

(c) noise associated with transit-time effects at high frequencies, and noise in the grid loading impedance, to which the influence of the grid-cathode capacitance, in conjunction with the cathode lead inductance, is equivalent.

(a) *Thermal Noise*—The noise power N_0 generated in a resistance in the frequency range f to $f + \Delta f$ is given by

$$N_0 = 4kT\Delta f \qquad (5.22)$$

where T is the absolute temperature and k is BOLTZMANN's constant. The corresponding r.m.s. noise e.m.f. developed in a resistance R is accordingly

$$E_n = \sqrt{N_0 R} = 1 \cdot 28 \times 10^{-10} \sqrt{R \Delta f} \qquad (5.23)$$

where R is measured in ohms, f in cycles per second, E_n in volts and we have put $T = 300°$ K.

In a distributed amplifier the noise voltage due to the output impedance of the source feeding the grid line, assumed equal to Z_{0g}, is amplified in the normal manner. The forward voltage wave in the anode line has amplitude

$$\frac{ng_m Z_{0a}}{2} \frac{\sqrt{N_0 Z_{0g}}}{2}$$

and the output power in the anode line developed across a matching termination Z_{0a} is accordingly equal to

$$\tfrac{1}{8} N_0 n^2 g_m^2 Z_{0a} Z_{0g} \qquad (5.24)$$

The terminating resistance in the grid line gives rise to a voltage wave travelling in the reverse direction in this line; such signals are

---

* See the companion volume in the present series: *Signal, Noise and Resolution in Nuclear Counter Amplifiers*, by A. B. GILLESPIE.

only amplified appreciably if the phase change per section is either approximately zero or approximately equal to π. The above expression for the output noise power may therefore have a quantity zeta times as much added to it where ζ approaches the value unity at low frequencies and at the line cut-off frequency and is small over most of the range.

A further contribution of $N_0/2$ is given by the reverse anode termination and the output matching termination together. The total output thermal noise power is thus equal to

$$\frac{N_0}{2}[1 + n^2 g_m^2 Z_{0a} Z_{0g}(1 + \zeta)/4] \tag{5.25}$$

(b) *Shot Noise*—The effect of shot and partition noise in a valve is usually expressed in terms of an equivalent resistance noise generator R_s in the grid circuit. The forward travelling noise power generated at each anode is

$$\tfrac{1}{4} N_0 g_m^2 Z_{0a} R_s$$

Since the contributions from the various valves are incoherent the total output noise power due to this cause is equal to

$$\tfrac{1}{4} N_0 n g_m^2 Z_{0a} R_s \tag{5.26}$$

(c) *Valve Input Noise*—The third source of noise only becomes important at the higher frequencies at which the valve input resistance approaches the value Z_{0g}. Attenuation of the desired signal then occurs in the grid line and such frequencies barely fall within the working range of the amplifier. This source of noise is of little practical significance.

The total output noise power is given by the sum of the contributions 5.25 and 5.26 above:

$$\frac{N_0}{2}[1 + n^2 g_m^2 Z_{0a} Z_{0g}(1 + \zeta)/4 + n g_m^2 Z_{0a} R_s/2] \tag{5.27}$$

and the total r.m.s. output noise voltage is given by

$$\sqrt{\frac{N_0 Z_{0a}}{2}[1 + n^2 g_m^2 Z_{0a} Z_{0g}(1 + \zeta)/4 + n g_m^2 Z_{0a} R_s/2]} \tag{5.28}$$

The equivalent noise signal input voltage is obtained by dividing this expression by the stage gain $n g_m Z_{0a}/2$.

It is seen that the effect of the resistive terminations in the anode line will ordinarily be negligible and that the contribution due to shot noise can be made small by employing a considerable number of valves in the stage. Attenuation in the grid line, however, and the possibility of noise arising from source (c) above, sets an upper limit to the number of valves which may usefully be employed.

The noise factor of the amplifier may be defined as the ratio of the total output noise power to the noise power at the output due solely to the amplified input noise. The factor is given by the following expression

$$1 + \zeta + \frac{4}{n^2 g_m^2 Z_{0a} Z_{0g}} + \frac{2R_s}{n Z_{0g}} \qquad (5.29)$$

A noise factor of about three and an equivalent noise signal input voltage of the order of 40 μV are estimated for the case of an amplifier employing six small valves to give a gain of three over a bandwidth of 200 Mc/s at an impedance level of 200 Ω.

Some complete distributed amplifiers described in the literature

Source	Bandwidth Mc/s	Voltage gain	Impedance Ohms		Valves		Maximum Output
			In	Out	No.	Type	
Copson [521]	30	2·5	200	200	3	6AC7	85 V r.m.s.
	30	3·5	200	480	3	807	
Cormack [522]	170	25	75	340	8	Z77	
Hewlett-Packard Type 460A	140	10	200	200			5 V peak
Kelley [523]	100	25	195	low	8	6AK5	
	90	40	high	470	16	6AK5	$\pm$ 60 V
Myers [525]	100	100	100	250	12	4X150A	$\pm$ 160 V
Rudenberg and Kennedy [526]	200	3·1	200	200	6	6AK5	
S.K.L. Type 204	200	100	200	200	24	6AK5	$\pm$ 6 V
Scharfman [527]	10–360	2·8	50	93	9	6AK5	
Tyminski [528]	250	2·8	180	180	6		
Weber [520]	3–400	2·8	150	220	5	6AK5	
Yu [529]	150	5000	200	200	58	various	130 V peak to peak push pull

5.4.3 Practical Circuits.

Theoretical and constructional details of a number of complete amplifiers are to be found in papers by COPSON [521], CORMACK [522], KELLEY [523], KENNEDY and RUDENBERG [524], MYERS [525], RUDENBERG and KENNEDY [526], SCHARFMAN [527], TYMINSKI [528], YU et al. [529]. Some data on these amplifiers are summarized in the table and circuit diagrams of two of them will be reproduced here by way of example.

5.4.3.1 Signal Amplifier for Cathode Ray Oscilloscope

An amplifier has been designed by KELLEY to provide a push-pull output for symmetrical deflection of a cathode ray tube up to a frequency of about 100 Mc/s. The complete arrangement consists of a probe unit, a pre-amplifier and a main amplifier.

(a) *Pre-amplifier*—The circuit is as shown in fig. 5.12. The probe unit V_1 has a high input impedance and plug-in capacity type attenuators may be used to prevent the valve from overloading. The output from the probe is fed into a ladder attenuator, the input impedance of which (195 Ω) provides the effective anode load for the valve V_1. A maximum attenuation of 20 db is provided in steps of 4 db.

The signal leaving the attenuator enters the grid line of the first distributed stage (valves V_2, V_3, V_4) which has a gain of about 5 times and a design cut-off frequency of 120 Mc/s. The anode line of this stage is coupled to a second similar stage (valves V_6, V_7, V_8) via a buffer valve V_5 which effectively provides an open circuit termination to the anode line, thus giving a factor of two improvement in gain. The output from the second stage is taken through a second buffer valve V_9 and is fed, from the anode of this valve, into a synchronizing amplifier (which triggers the cathode ray tube time base circuit) and also via a length of 200 Ω delay line (see § 6.3.1) into the main amplifier.

Filter sections of the m-derived type are employed throughout and the method of construction* is indicated in fig. 5.13.

---

* Several workers have used self-supporting continuous helices, of thick wire, so designed that the coefficient of coupling between adjacent sections has the correct value. It is felt, however, that better mechanical stability is obtained if thinner wire is employed and the coils are wound on a threaded insulated former.

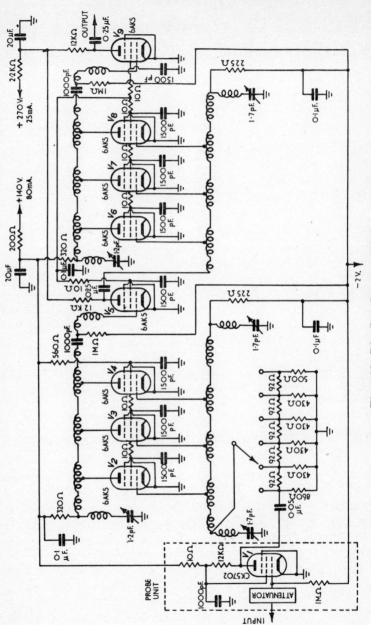

Fig. 5.12. Distributed pre-amplifier.

(b) *Main Amplifier*—The main amplifier, depicted in fig. 5.14, was designed to have a pass band of about 90 Mc/s; a voltage swing of at least 60 V, in either direction, was desired at the output. The input delay line is terminated by a 200 Ω carbon resistor of small physical size. A small inductance is connected in series with this resistor and is very effective in compensating for the shunt capacitance of the load. The inverter valve V_1 was proved to be

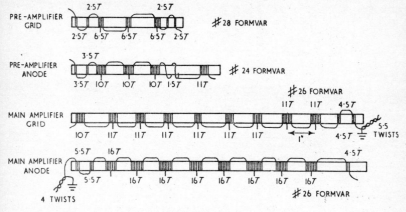

Fig. 5.13. Delay line construction.

completely satisfactory within the frequency band involved and did not introduce any noticeable deterioration in output pulse rise-time. Two valves V_2 and V_3 follow the inverter and drive the grid lines of the push-pull output stages. Here again the grid line impedances form the load resistances of the coupling valves. The lines are unterminated at their input ends; no matching at the inputs is required if the remote ends are sufficiently well terminated and an attenuation factor of two is thereby avoided. The anode lines are connected to the cathode ray tube deflector plates without the addition of any resistive termination. A factor two increase in gain once more results, but adjustment of the reverse termination is more critical.

Satisfactory performance was obtained when the screening grids of each stage were simply connected in parallel and decoupled to earth by a single capacitor in the middle of the stage. Shielding between the anode and grid lines was found to be necessary in the main amplifier but not in the pre-amplifier.

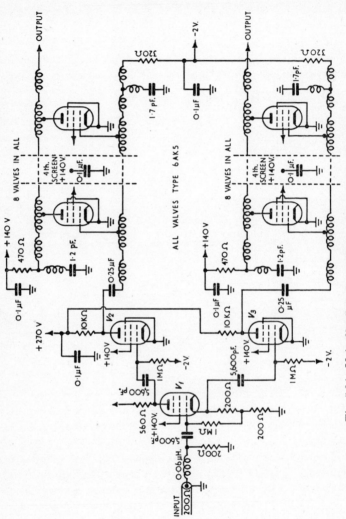

Fig. 5.14. Main distributed amplifier giving a push-pull output.

Constructional details of the delay lines are given in fig. 5.13. The small capacitances of 1·2 pF and 1·7 pF, on the anode and grid lines respectively, are obtained by twisting together an earth lead with a short length of wire which forms part of the coil winding. Dissipation in the grid line (due to the input conductance of the valves), and the unavoidable shunt capacitance of the termination, require that an inductive component be inserted in series with the grid terminating resistor. An inductance of 0·1 μH was found to be suitable; this may be constructed by winding 11 turns of No. 26 wire (No. 28 S.W.G.) on a 0·1 in. diameter insulated former.

The main amplifier alone was found to have a gain of about 40 times and the output pulse had a rise-time of 5 mμsec. Figures of 1000 for the gain and 6·3 mμsec. for the rise-time were obtained when the main amplifier and pre-amplifier were combined.

5.4.3.2 An Amplifier of 400 Mc/s Bandwidth

Fig. 5.15 shows the circuit diagram* of an amplifier which provides a voltage gain of 2·8 over a frequency band extending from a few megacycles per second up to about 400 Mc/s. Filter sections of the m-derived type were employed for the grid and anode lines. The former had a characteristic impedance of 150 Ω and each coil had a total inductance of 0·138 μH. The inductance of each half was 0·049 μH and the mutual inductance between sections had a value of 0·020 μH. Corresponding figures for the anode line, of 220 Ω impedance, were 0·167 μH, 0·059 μH and 0·024 μH respectively. Some difficulty was encountered in obtaining the required coefficient of coupling between adjacent sections and both grid and anode coils, of the self supporting type, were wound with the end turns placed "inside" the rest of the coil. Absolute minimum lengths of connecting lead were employed in the terminating networks.

The valve input conductance was lowered by employing the technique outlined in § 5.4.2.3. The cathode "line" consisted of two size 16 wires (No. 18 S.W.G.), in parallel, spaced ¼ in. apart. The compensating screen series inductance was obtained by utilizing the inductances of the ½ in. long lead to the screen bypass capacitor, the optimum value being determined experimentally. No attempt was

---

* The authors are endebted to Professor J. WEBER, University of Maryland, in association with U.S. Naval Ordance Laboratory, Silver Spring, Md., for details of this amplifier. The work was presented at the 1951 Institute of Radio Engineers Convention.

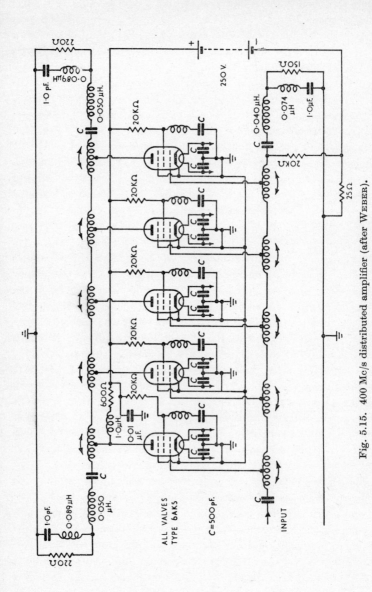

Fig. 5.15. 400 Mc/s distributed amplifier (after WEBER).

made to screen the grid line from the plate line nor were any valve shields employed.

An output pulse rise-time of less than 2 mμsec. was measured.

5.5 SPECIAL VALVES

5.5.1 Transmission-Line Tubes.

Some consideration may now be given to the possibility of obtaining wide-band amplification at frequencies above 400 Mc/s by the use of special tubes involving transmission line concepts. PERCIVAL's original idea was to wind the anode and grid lines in the form of uniform helices, arranged concentrically, surrounding a long straight cylindrical cathode. Cylindrical screening and suppressor grids may be incorporated and the whole is enclosed in an evacuated envelope with leads in at both ends. Electrons move radially from the cathode, through the control and other grids, to the anode, in the normal manner, and the tube would behave exactly as the distributed amplifiers described above if coupling between the coils is neglected. Improved high frequency performance is expected for two principal reasons (i) the line parameters are truly distributed and (ii) no question of cathode lead inductance can arise since the cathode itself forms one conductor of the transmission system.

An analysis, using circuit theory, has been made by LEWIS [530], in which inductive and capacitative coupling between the anode and grid coils is taken into account. The effect of the electrons is allowed for by introducing a mutual shunt conductance, per unit length, between the two helices and by including a distributed shunt resistance between each coil and earth (cathode). It is found that waves may travel with two velocities which are, in general, different. In the faster mode, the grid and anode voltage signals are in phase and in the slower mode these voltages are in opposite phase. The characteristic impedances associated with the two modes are different, in one and the same line, but reduce to the same value, which is resistive and independent of frequency, if the coils are wound in opposite directions (in a concentric system) and if the coefficients of inductive and capacitative coupling are made equal. Complete matching of the tube is then possible, at least in principle.

It is found that signals travelling in either direction in the antiphase mode are amplified according to an exponential law with length and waves in the in-phase mode are attenuated according to the same law. Signals in each mode appear on both the anode and grid coils simultaneously and are amplified (or attenuated) to the same extent in the two lines. Advantages which would be possessed by the tube include (i) much greater gain for the same H.T. supply power, (ii) the availability of a push-pull output from a single-ended input, and (iii) the bilateral nature of the device, which suggests its application as a broad-band repeater amplifier in a two-way communication link.

Performance would be limited at high frequencies by phase distortion in the helical lines and by transit-time effects. Extremely close grid-cathode spacing would be required with consequent difficulties in construction, bearing in mind the considerable length of the tube. The valve will tend to oscillate if the product of the reflection coefficients at the two ends times the gain for the double journey exceeds unity. The proper matching of the tube over a wide frequency band presents a major problem and close tolerance construction will be necessary in order to avoid reflections from irregularities along the length of the tube. Analysis also reveals the existence of a low frequency limit to the uniform amplification band, a property which would be disadvantageous in some applications.

There remains the alternative of constructing a tube in which coupling between the anode and grid lines is sedulously avoided. The possibility of oscillation due to imperfect matching will be removed and constructional tolerances may be relaxed, further there will be no low frequency limit to the uniform amplification band. The close grid-cathode spacing must be maintained, however, and no saving in supply power is achieved. Active theoretical and experimental work on the coupled type of tube is being carried out by FOWLER [531], [532].*

5.5.2 Travelling-Wave Tubes.

Much is now known concerning the theory and performance of travelling wave tubes of the KOMPFNER-PIERCE [533] and allied types. (Also see MATHEWS [534].) These provide high gain and large band-widths but possess a lower frequency limit of the order of 1000 Mc/s. Such valves accordingly belong to microwave modulated carrier systems, and there does not appear to be any immediate

---

* For additional information on coupled systems see references given in § 3.6.

possibility of their range of application being extended down into the millimicrosecond pulse region.

This type of tube employs a linear beam of electrons which travel at high velocity in a straight line; interaction occurs between the beam and signals propagated on a slow-wave structure placed in close proximity to it. Input and output terminals are at the two ends of the slow-wave structure but the tendency to oscillation due to mismatch can be counteracted by the introduction of attenuation. The attenuation introduced affects both the incident and reflected waves and thus has a double effect in preventing oscillation but only a single effect on the wanted output. High gain and stability are therefore not incompatible, as in the case of the bilateral coupled transmission line tube.

Further developments (ADLER [535]) will undoubtedly take place in the attempt to bridge the gap between the distributed amplifier employing conventional valves, and the microwave travelling wave tube.

6

CATHODE RAY OSCILLOSCOPES

6.1 INTRODUCTION

THE cathode ray oscilloscope needs little introduction to the reader. It is indispensable to the research worker using millimicrosecond pulse techniques who will wish to understand the exact operation of his circuits, and to examine the waveforms of the pulses present in the system. Again, the cathode ray oscilloscope is often a convenient, and probably the most reliable, instrument for the measurement of time intervals between pulses, their relative amplitudes and their shapes. In particular, if the pulses to be measured are of a transient nature, the oscilloscope fitted with a photographic recording camera is the only available instrument by means of which an exact analysis of the pulse waveforms may be made. The need for careful examination of pulse waveforms is of special importance when using very short duration pulses, since the presence of stray lead inductances and capacitances may easily cause unexpectedly serious distortions, thus giving rise to errors in the measurement of amplitudes and time intervals.

The design of cathode ray oscilloscopes suitable for the display and recording of single transient pulses will first be considered, and then the oscilloscope display of regular recurrent pulses will be described. It turns out that this latter case presents a rather easier problem than does the former. In both cases the major problem lies in the design of suitable cathode ray tubes and therefore these tubes will be described first. The associated time-base and other circuits will likewise be discussed, and the chapter concludes with a description of the special circuit technique of pulse sampling which may be used to display regular recurrent pulses.

6.2 CATHODE RAY TUBE DESIGN

Several factors in the design of a cathode ray tube become of special importance when very short duration transient pulses are to be displayed.

6.2.1 Transit-Time Limitations.

In a conventional type of cathode ray tube the dimensions of the plates and the electron beam velocity are such that the transit-time of the beam through the deflection plate system is in the region 1 to 10 mμsec. If the deflection voltage changes appreciably during this interval of time, then the resultant pattern displayed on the tube screen will not be a true representation of the deflecting voltage waveform.

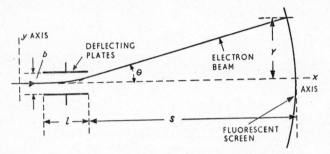

Fig. 6.1. Electron beam deflection in cathode ray tube.

We shall illustrate this effect by considering the application of two simple types of voltage waveform to a pair of parallel deflecting plates, through which an electron beam is passing (fig. 6.1). The problem may be simplified by assuming that the electrostatic field acting on the electron beam, due to the potential difference between the deflecting plates, is uniform over the length of the plates and is negligible beyond their ends. We shall further suppose that the diameter of the beam is small compared with the distance between the plates.

For the motion of a single electron through the deflecting plate system, we have:

$$\text{axial electron velocity} = \frac{dx}{dt} = v_x \quad \text{(independent of deflecting voltage)} \tag{6.1}$$

and

$$m\frac{d^2y}{dt^2} = \frac{E}{b}e \quad \text{where } e \text{ and } m \text{ are the charge and mass of an electron.} \tag{6.2}$$

6.2.1.1. Pulse Conditions

Let us first consider the case of a step function of voltage, of amplitude E, applied at time $t = 0$, and suppose that a particular electron enters and leaves the deflecting plate system at times t_1 and t_2 respectively.

When the electron emerges from the deflecting field, having been given a transverse velocity v_y, equation 6.2 yields:

$$v_y = \int_{t_1}^{t_2} \frac{E}{b} \cdot \frac{e}{m} \, dt \qquad (6.3)$$

$$= \frac{E}{b} \cdot \frac{e}{m} t_r$$

where $t_r = t_2 - t_1$ is the electron transit time through the deflecting plates.

If the electron is between the deflecting plates when the step voltage is applied (i.e., $t_1 < 0$) then on emerging from the deflecting field,

$$v_y = \int_0^{t_2} \frac{E}{b} \cdot \frac{e}{m} \, dt$$

$$= \frac{E}{b} \cdot \frac{e}{m} t_2 \qquad (6.4)$$

The resultant deflection Y on the screen of the cathode ray tube for small values of θ is given by:

$$Y = S\theta$$
$$= Sv_y/v_x \qquad (6.5)$$

The complete solution for Y, expressed as a function of the time t after the application of the step voltage, is then,

$$Y = 0 \qquad \text{for } t < \tau$$

$$Y = \frac{E}{b} \cdot \frac{e}{m} \cdot \frac{S}{v_x} \cdot (t - \tau) \qquad \text{for } \tau < t < (\tau + t_r)$$

$$Y = \frac{E}{b} \cdot \frac{e}{m} \cdot \frac{S}{v_x} \cdot t_r = Y_0 \quad \text{say, for } t > \tau + t_r$$

where τ is the time of flight of the electron from the deflecting plates to the screen.

The corresponding waveforms are shown in fig. 6.2 and illustrate

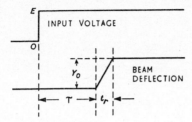

Fig. 6.2. Transit time distortion of display of step function.

that the electron transit time t_r sets a lower limit to the rise-time of a pulse that can be displayed.

6.2.1.2 Sine Wave Conditions

Let the applied voltage take the form

$$E = E_0 \sin \omega t$$

Then we have from equation 6.3

$$v_y = 2 \frac{E_0}{\omega b} \cdot \frac{e}{m} \sin \tfrac{1}{2}\omega(t_1 + t_2) \sin \tfrac{1}{2}\omega t_r$$

The deflection Y on the screen of the cathode ray tube is then,

$$Y = 2 \frac{S}{v_x} \frac{E_0}{\omega b} \frac{e}{m} \sin \tfrac{1}{2}\omega(t_1 + t_2) \sin \tfrac{1}{2}\omega t_r$$

This deflection is a sine wave of the deflecting waveform frequency and peak amplitude given by:

$$Y_0(\omega) = E_0 \frac{2Se}{\omega v_x b m} \sin \tfrac{1}{2}\omega t_r$$

When

$$\omega \to 0, \quad Y_0(\omega) \to E_0 \frac{Se}{v_x b m} t_r = Y_0(0) \text{ say}$$

hence

$$\frac{Y_0(\omega)}{Y_0(0)} = \frac{\sin(\tfrac{1}{2}\omega t_r)}{\tfrac{1}{2}\omega t_r} \tag{6.6}$$

i.e., the deflection sensitivity gradually diminishes to zero as the frequency is raised from a low value. At still higher frequencies the sensitivity rises again, as illustrated in fig. 6.3. Hence, if a cathode

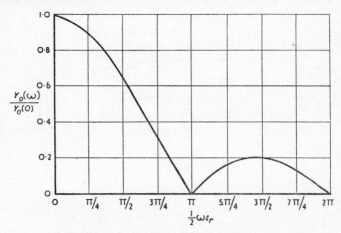

Fig. 6.3. Transit-time distortion of display of sine wave.

ray tube is being used to measure the amplitude of a high frequency sine wave, this loss of deflection sensitivity must be borne in mind in the amplitude calibration.

6.2.2 Connections to Deflecting Plates.

The inductance of the leads conveying the signal voltage to the deflecting plates must be kept to as low a value as possible if the

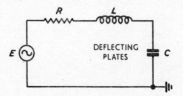

Fig. 6.4. Equivalent circuit of deflecting plates.

cathode ray tube is to be used for displaying fast rising pulses or high frequency sinusoidal waveforms. The normal cathode ray tube practice of bringing these connecting leads through the base of the tube is quite inadmissable for the present applications and the leads should be sealed through the glass envelope adjacent to the deflecting plates. The following simple analysis of the circuit will show that the effect of lead inductance is as important as the transit-time limitation in determining the shortest pulse rise time or highest frequency sine wave that can be displayed without undue distortion.

The simple circuit representing the deflecting plate system is shown in fig. 6.4, where R is the total resistance of the external circuit, including that of the deflecting voltage generator.

6.2.2.1 *Pulse Conditions*

Let a step voltage,* of amplitude E, be applied to the circuit, then the Laplace transform of the voltage appearing between the deflecting plates is given by,

$$\bar{V} = \frac{E}{p(p^2 LC + pCR + 1)}$$

The solution of this equation (see Table 1.1, page 10), when R is less than the critical damping resistance, is:

$$V = E \left[1 - e^{-\alpha t} \left(\frac{\alpha}{\beta} \sin \beta t + \cos \beta t \right) \right] \qquad (6.7)$$

where

$$\alpha = \frac{R}{2L}$$

$$\beta = \sqrt{\frac{1}{LC} - \frac{R^2}{4L^2}}$$

In a practical circuit we choose R such that *the resultant voltage across the deflecting plates has as short a time of rise as possible, without an undue amount of overshoot above the amplitude E of the input voltage*. A good compromise is to make $\frac{\alpha}{\beta}$ unity which gives an overshoot of approximately 4%. If

$$\frac{\alpha}{\beta} = 1 \quad \text{then} \quad R = \sqrt{\frac{2L}{C}}$$

---

* This type of input is considered before the sinusoidal case since we are interested in pulses rather than in simple harmonic signals. We also wish to establish a suitable value for the damping resistance on transient response before analysing the sinusoidal case.

Putting
$$f_0 = \frac{1}{2\pi\sqrt{LC}}, \text{ the circuit resonant frequency,}$$
we have
$$R = \frac{1}{\sqrt{2}\pi f_0 C} = \frac{0\cdot 22}{f_0 C} \tag{6.8}$$

This latter formula may be used for estimating the value of R required; both C and f_0 may be measured for a particular cathode ray tube.

With this value of damping resistance
$$\alpha = \beta = \frac{R}{2L} = \frac{1}{CR} = \frac{1}{T} \quad \text{where} \quad T = CR$$

Thus the deflecting voltage becomes
$$V = E\left[1 - e^{-t/T}\left(\cos\frac{t}{T} + \sin\frac{t}{T}\right)\right] \tag{6.9}$$

At the instant when $V = E$, on the rising edge of the pulse, we have
$$\cos\frac{t}{T} + \sin\frac{t}{T} = 0$$

Therefore
$$t = \frac{3\pi}{4}T = 2\cdot 3T \tag{6.10}$$

This gives a measure of the rise-time of the voltage waveform displayed on the cathode ray tube screen. The shape of the curve is given in fig. 6.5 ($t_r/T = 0$).

6.2.2.2. Sine Wave Conditions

Let the input waveform take the form of a sine wave of real amplitude E_0. The resultant complex voltage amplitude across the deflecting plates is given by:
$$\bar{V}_0 = \frac{1}{j\omega CR + (1 - \omega^2 LC)}E_0$$

If we now use the same value of damping resistance R, as obtained in the preceding pulse analysis, we obtain for the peak value of the deflecting voltage
$$V_0(\omega) = E_0(1 + \tfrac{1}{4}\omega^4 T^4)^{-\frac{1}{2}} \tag{6.11}$$

where we have written $V_0(\omega)$ to indicate the dependence of V_0 on frequency. The resultant amplitude of the displayed waveform is proportional to this value and is of sinusoidal form with a phase shift between the input voltage and the displayed waveform.

At low frequencies $V_0(\omega) \to E_0$.
Hence

$$\frac{V_0(\omega)}{V_0(0)} = \frac{Y_0(\omega)}{Y_0(0)} = (1 + \tfrac{1}{4}\omega^4 T^4)^{-\frac{1}{2}} \qquad (6.12)$$

This expression is plotted in fig. 6.6 ($t_r/T = 0$), and illustrates how the deflection sensitivity of the cathode ray tube diminishes as the frequency is raised, owing to the inductance and capacitance in the deflecting plate circuit.

6.2.3 Overall Frequency Limitations of Normal Deflecting Plate Structures.

A simple analysis of the effects of transit-time and deflecting plate lead inductance on the rise-time of an input step voltage or on a sinusoidal voltage waveform has been given in the preceding two sections. In a practical cathode ray tube both of these effects occur so that it is of importance to combine the two results. It should be noted that in the design of the cathode ray tube these effects are independent.

6.2.3.1 Pulse Conditions

The voltage appearing across the plates inside the cathode ray tube is given by equation 6.9, and inserting this value in equation 6.3 we obtain:

$$v = \int_{t_1}^{t_2} \frac{E}{b} \cdot \frac{e}{m} \left[1 - e^{-t/T} \left(\cos\frac{t}{T} + \sin\frac{t}{T} \right) \right] dt$$

Performing the integration we find:

$$v_y = \frac{E}{b} \cdot \frac{e}{m} \left[t + T e^{-t/T} \cos\frac{t}{T} \right]_{t_1}^{t_2}$$

The resultant deflection on the screen of the cathode ray tube is then given by

$$Y = S v_y / v_x$$
$$= \frac{E}{b} \cdot \frac{e}{m} \cdot \frac{S}{v_x} \left[t + T e^{-t/T} \cos\frac{t}{T} \right]_{t_1}^{t_2} \qquad (6.13)$$

If the electron is situated between the deflecting plates when the step voltage is applied, the limit of integration t_1 in this equation is replaced by zero.

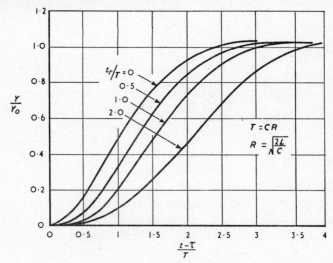

Fig. 6.5. Transit-time and circuit distortion of display of unit function.

(t_r = transit-time between plates.)
(τ = time of flight from plates to screen.)

The complete expression for the deflection Y expressed as a function of the time t after the application of the step voltage is then

$$Y = 0 \qquad \text{for } t < \tau$$

$$Y = \frac{E}{b} \cdot \frac{e}{m} \cdot \frac{S}{v_x} \left[(t - \tau) - T + Te^{-\frac{(t-\tau)}{T}} \cos \frac{t-\tau}{T} \right]$$
$$\text{for } \tau < t < (\tau + t_r)$$

$$Y = \frac{E}{b} \cdot \frac{e}{m} \cdot \frac{S}{v_x} \left[t_r + Te^{-\frac{(t-\tau)}{T}} \left\{ \cos \frac{t-\tau}{T} - e^{\frac{t_r}{T}} \cos \frac{(t - \tau - t_r)}{T} \right\} \right]$$
$$\text{for } t > (\tau + t_r)$$

The waveform appearing on the screen of the cathode ray tube for this pulse condition is shown in fig. 6.5; curves are plotted for different values of the transit-time.

6.2.3.2 Sine Wave Conditions

Combining equations 6.11 and 6.6 we obtain:

$$\frac{Y_0(\omega)}{Y_0(0)} = \frac{\sin(\tfrac{1}{2}\omega t_r)}{\tfrac{1}{2}\omega t_r}(1 + \tfrac{1}{4}\omega^4 T^4)^{-\tfrac{1}{2}} \qquad (6.14)$$

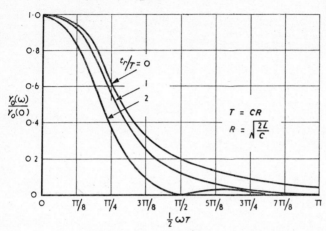

Fig. 6.6. Transit-time and circuit distortion of display of sine wave.

Curves for different values of the transit-time t_r are plotted in fig. 6.6. The frequency characteristic of the deflecting plate structure of a particular cathode ray tube and the shortest possible rise time of any displayed pulse may be determined by means of figs. 6.5 and 6.6. The practical effects of transit-time and deflecting plate lead inductance may be seen from the following numerical values which were measured on a cathode ray tube designed for the observation of transient millimicrosecond pulses.

Deflecting plate capacitance $C = 6$ pF
Resonant frequency f_0 $= 250$ Mc/s
Transit-time t_r $= 1\cdot 3$ mμsec. (estimated)
$T = CR = 1$ mμsec. (equation 6.8)
Required damping resistance $R = 160\ \Omega$

Thus T and t_r are similar in value and are of equal importance in limiting the minimum pulse rise-time. For this particular tube the minimum rise-time of the displayed waveform, for a step voltage input, is about 2 mμsec. measured between 10% and 90% amplitude points. In the case of a sine wave input the deflection sensitivity at a frequency of 200 Mc/s is 0·7 of its low frequency value.

6.2.3.3 Time-Base Waveform

A further important case to consider is the application of a time-base waveform to the deflecting plates.

Let the input voltage waveform, derived from the time-base circuit, be:

$$E = 0 \qquad\qquad t < 0$$
$$E = Kt \quad \text{where } K \text{ is a constant}, \quad t \geq 0$$

Then the voltage V applied between the deflecting plates inside the cathode ray tube is given by

$$\bar{V} = \frac{K}{p^2(p^2 LC + pCR + 1)}$$

Taking the inverse transformation, we have:

$$V = K\left[(t - T) + e^{-\alpha t}\left\{\frac{\alpha T - 1}{\beta}\sin \beta t + T \cos \beta t\right\}\right] \quad (6.15)$$

where

$$\alpha = \frac{R}{2L}$$

$$\beta = \sqrt{\frac{1}{LC} - \frac{R^2}{4L^2}}$$

$$T = CR$$

and

$$V = 0, \frac{dV}{dt} = 0 \text{ at } t = 0$$

If we again choose R to give a slightly underdamped circuit as in § 6.2.2.1, the solution 6.15 reduces to

$$V = K\left[t - T\left\{1 - e^{-t/T}\cos\frac{t}{T}\right\}\right]$$

The effect of transit-time distortion is included by inserting this expression for the voltage into equation 6.3. Performing the integration, we obtain:

$$v_y = \frac{K}{b} \cdot \frac{e}{m}\left[\tfrac{1}{2}t^2 - Tt + \tfrac{1}{2}T^2 e^{-t/T}\left(\sin\frac{t}{T} - \cos\frac{t}{T}\right)\right]_{t_1}^{t_2} \quad (6.16)$$

$$\text{for } t_1 > 0$$

The resultant deflection D on the screen of the cathode ray tube in terms of the time t after the start of the applied time-base is then obtained as in § 6.2.1.1.

Fig. 6.7 shows the results in graphical form. From these curves it is seen that the chief effect of the deflecting plate circuit reactance and transit-time distortion is to give a time delay between the start of the time-base and the attainment of a steady scan velocity. The

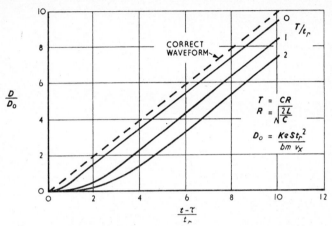

Fig. 6.7. Transit-time and circuit distortion of display of time-base waveform.

displayed time-base is still linear except for a short period at the commencement of the trace.

In a practical cathode ray tube, the time-base deflecting plate constants will be of the same order of magnitude as the signal deflecting plate constants; time-base distortions are accordingly insignificant since the time-base duration is generally at least ten times the shortest rise-time of the pulses to be displayed.

A simple geometrical arrangement has been assumed in the preceding analysis and the equivalent circuit has been simplified by the use of lumped constants. These assumptions, however, are justified in practice, and the results of the analysis give a useful indication of the frequency characteristics of the deflecting plate structure.

6.2.4 Methods of Reducing Display Distortions due to Deflecting Plates.

We have seen in the preceding section that both the electron transit-time and the deflection plate circuit reactances are limiting factors in the high frequency performance of a cathode ray tube. Hence, both of these factors must be reduced as far as possible if very short duration pulses are to be displayed. Two basic designs have been employed for this purpose, and will now be described.

6.2.4.1 Lee's Three Beam Micro-oscillograph

LEE [601] has designed an oscillograph tube in which the transit-time through the deflection plates was reduced by using a high beam velocity (50 KV accelerating voltage) and a short plate length. In addition, care was taken to reduce the plate lead inductances to a minimum. By this straightforward technique, the bandwidth was increased to 8600 Mc/s for 3 db attenuation. The disadvantage of this technique is that the deflection sensitivity is much reduced, as will be seen from equation 6.5 (the deflection is proportional to the transit-time and inversely proportional to the beam velocity). The difficulty was overcome by using VON ARDENNE's [602] technique of reducing the diameter of the electron beam so that a very small spot (0·01 mm diameter) was produced on the screen of the tube. The deflection sensitivity of a cathode ray tube may be expressed in terms of the deflection voltage required to move the beam by one spot diameter (see § 6.2.5); thus, by reducing the electron beam diameter, Lee was able to obtain a deflection sensitivity of 10 V per spot width. It is necessary to view the screen through a low power microscope because of the small physical size of the display.

6.2.4.2 Travelling-Wave Deflection Systems

Such methods are used to obtain a high deflection sensitivity whilst increasing the bandwidth of the deflecting system.

In a large number of applications, the amplitudes of the pulses to be examined are comparatively small, and some amplification must be included in order that the pulse displayed on the screen be sufficiently large for measurement purposes. The basic amplifier designs for this purpose have been described in Chapter 5. If a large output voltage is required for oscilloscope deflection, many valves must be employed and a large amount of D.C. power consumed; any method by which the tube sensitivity may be increased is accordingly extremely valuable. In applications where a large amplitude signal is available to drive the deflecting plates, however, the use of travelling wave systems is an unnecessary complication.

Reduction of electron transit-time and care in the design of the connection circuit are aimed at in the construction of a deflection system of large bandwidth. The former can obviously be achieved by shortening the length of the plates. Several pairs of short plates situated in tandem along the beam axis and suitably connected together may be employed in certain cases, in order to offset the consequent loss in deflection sensitivity. The length of the plates, connected as indicated in fig. 6.8, must be such that the transit-time

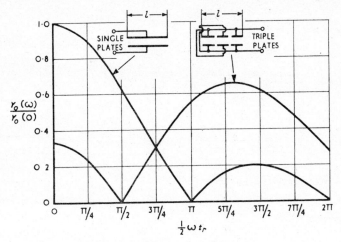

Fig. 6.8. Transit time distortion of display of sine wave for single and triple deflecting plates.

through one pair is approximately equal to half the period of the high frequency sine wave to be displayed. The deflections due to successive pairs of plates are then additive, and an overall sensitivity, equal to the number of plates multiplied by the sensitivity associated with one pair, is obtained. HOLLMAN [603], who has used this technique, gives the expression

$$\frac{Y_0(\omega)}{Y_0(0)} = \frac{\sin(\frac{1}{2}\omega t_r) - 2\sin(\frac{1}{6}\omega t_r)}{\frac{1}{2}\omega t_r} \quad (6.17)$$

for the variation of sensitivity with frequency for a triple system; here t_r is the total transit time for the three pairs of plates. This result is plotted in fig. 6.8, and may be compared with the curve for a single pair of plates (relation 6.6) of the same total length. It is seen that the arrangement gives a marked improvement but only over a range of frequencies round the design figure.

PIERCE [604], in his design of a travelling wave deflection system, has shown how the multiple plate technique may be modified so that a wide range of frequencies is covered; the tube then becomes suitable for displaying millimicrosecond pulses. The capacitances between the deflecting plates (fig. 6.9) together with small connecting inductances form a low pass filter network (§ 2.4). The constants of

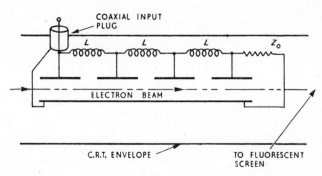

Fig. 6.9. Schematic diagram of travelling-wave deflecting system.

this network are so chosen that the velocity of propagation of a pulse is equal to the velocity of the electron beam. Ideally, each individual segment of the beam passing through the structure experiences the same deflecting field, so that there is no transit-time distortion. This would be the case if the length of deflecting plates was very small but, in practice, the shortest pulse rise-time that may be displayed is limited by the transit-time distortion of each of the deflecting plates, and by the frequency characteristic of the low pass filter network. The basic design equations are given in § 2.4.1, but a restatement of the formulae will help the reader to understand the operation of this deflecting plate system:

$$f_c = \frac{1}{\pi\sqrt{LC}}$$

$$T = \sqrt{LC}$$

$$Z_0 = \sqrt{\frac{L}{C}}$$

Here L and C are the inductance and capacitance per section of the filter (assumed to be of constant k type), f_c is the cut-off frequency, T the time delay per section, and Z_0 the characteristic impedance.

From these equations we have:

$$f_c = \frac{1}{\pi T} \quad (6.18)$$

$$T = CZ_0 \quad (6.19)$$

From equation 6.6, fig. 6.3, we see that the sensitivity falls, due to the effect of transit-time, to a value,

$$\frac{Y_0(\omega)}{Y_0(0)} = 0{\cdot}7$$

when $\tfrac{1}{2}\omega t_r \simeq 1{\cdot}5$. This value corresponds to a frequency

$$f = \frac{1{\cdot}5}{\pi t_r} \quad (6.20)$$

The system is designed such that $T \simeq t_r$. Thus, comparing equations 6.18 and 6.20, we see that the reduction of deflection sensitivity at high frequencies, due to the electron transit-time, through one pair of plates will always be rather less serious than that due to the frequency cut-off effect of the deflecting plate structure.

Equation 6.18 shows that in order to increase the cut-off frequency of the deflecting plate structure it is necessary to reduce the time delay per section. This means that either the electron velocity must be increased, or the deflection plates reduced in size. Both of these procedures result in a reduction of deflection sensitivity, or alternatively, more deflection plates are required for a given sensitivity. Thus, in a practical design of tube, the cut-off frequency is kept low, consistent with the requirements of the waveforms to be displayed, so that as great a deflection sensitivity as possible may be obtained with minimum complexity. Having determined what cut-off frequency is required, the time delay per section T is fixed so that for a given impedance Z_0 the capacitance C per section may be determined. The characteristic impedance Z_0 is normally chosen to match a particular coaxial cable impedance and the connections to the deflection plate structure are made via coaxial plugs or sockets sealed through the glass wall of the tube.

The cathode ray tube described by PIERCE was designed to have as large a deflection sensitivity as possible, consistent with a reasonable bandwidth, and the actual performance figures were as follows:

Cut-off frequency f_c	1200 Mc/s for 3 db attenuation
Characteristic impedance Z_0	75 Ω
Electron gun acceleration voltage	2000 V
Screen spot diameter	0·02 mm
Deflection sensitivity	0·027 V per spot width

A low value of beam current was used in order to obtain a very small diameter spot. Only regular recurrent waveforms could therefore be displayed, owing to the lack of brightness.

In this type of deflection system, it is necessary to terminate the filter network by a resistance equal to its characteristic impedance. If this is not done, the signal travelling down the filter will be reflected back along the network, so producing a spurious deflection on the screen of the tube. The terminating resistance may be inserted inside the tube envelope.

6.2.4.3 Travelling-Wave Deflecting Plates with Distributed Constants

Discussion in the previous section has been concerned with the case where the connections between pairs of deflecting plates were in the form of lumped inductances. It is a simple step from this filter network to arrange the deflection system in the form of a transmission line, such as two parallel flat strips, with spacing and width to give the required characteristic impedance. The speed at which a pulse travels along the transmission line is much greater than that of the electron beam; the line must be accordingly coiled up in some manner so that the path traversed by the pulse is considerably longer than that followed by the electrons. The line is so designed that the resultant velocity of transmission of a pulse, in the direction of the electron beam, is the same as the beam velocity. Transit-time distortion is thus eliminated, except for that due to the finite width of the strips forming the transmission line.

This technique has been employed by several research workers (see OWAKI et al. [605]), and fig. 6.10 shows the deflecting plate structure designed by SMITH [606] and his colleagues. The line consists of a metal strip, about 1 cm wide, wound into a five turn helix of radius 1 cm, flattened on one side. The helix is mounted inside an earthed shield with a spacing of approximately 1½ mm between the helix and shield, so forming an unbalanced line of 50 Ω characteristic impedance. Connections are made to the helix at each end by 50 Ω coaxial plugs sealed through the glass wall of the cathode ray tube. The electron beam, accelerated by a potential of 10 KV, passes between the flattened part of the helix and the adjacent earthed shield. The perimeter length of the helix was arranged such that the beam and pulse velocities were equal. The beam transit

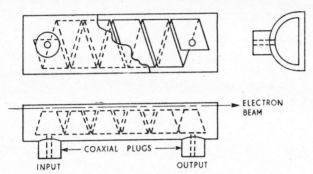

Fig. 6.10. Distributed travelling wave deflection structure.

time per turn of the helix amounted to 0·17 mμsec., resulting in a 4% reduction in deflection sensitivity at a frequency of 940 Mc/s.

This tube was designed for the photographic recording of transient pulses, and accordingly the beam current and accelerating potentials were chosen to give the necessary trace brightness. The following figures were obtained:

Deflection sensitivity = 0·6 V per spot width

Spot size = 0·15 mm

Total accelerating potential = 35 KV

Light output sufficient to record on film a writing speed of 10^{11} spot widths per second.

The deflection sensitivity obtained should be compared with the figure given in § 6.2.4.1 for LEE's tube.

6.2.5 Spot Size and Deflection Sensitivity.

In normal low speed cathode ray oscilloscopes, it has been customary to consider the deflection sensitivity in terms of the voltage required to deflect the spot on the screen through a given distance, usually 1 cm. The screen display may be magnified or diminished, however, by optical methods, so this definition of deflection sensitivity has no fundamental significance. In the case of oscilloscopes for displaying short duration pulses, the amplitude of the signal pulse may be small, and since the provision of suitable signal amplifiers with the necessary bandwidth is difficult and costly, it becomes important to design the tube for maximum deflection sensitivity. The smallest amplitude pulse that can be detected on the screen, magnified by optical methods if necessary, will be limited by the spot width, and is usually taken as being that pulse which gives one spot width deflection.* Hence it is preferable to define the deflection sensitivity of a tube in terms of the signal voltage required to produce such a deflection. For a given cathode ray tube deflection plate system, the sensitivity may be increased by reducing the spot size. The latter is usually accomplished by lowering the electron beam current, but there is a consequent loss of brightness at the screen. A good example of this technique is found in PIERCE's travelling wave tube, described in the previous Section, in which the spot width was reduced to the very small size of 0·001 in.; a signal pulse giving a deflection of 0·01 in. can then be considered as providing a good display for measurement. In this case, the displayed pulse is so small that a low power microscope must be used for visual observation and measurements.

From this discussion of deflection sensitivity, it is at once obvious that the electron optics of a cathode ray tube must be very carefully designed to give the smallest spot width for a given electron beam current. The term 'spot width' has been used in a rather general sense heretofore, but, if necessary, it may be precisely defined by reference to the curve of spot brightness versus distance across a spot diameter. Fig. 6.11 shows a representative curve taken by moving a very narrow slit across the spot and measuring the brightness at different positions. The spot width may then be defined as the width of the curve between half peak intensity points. The brightness curve only falls to zero at a considerable distance from the centre of the spot owing to internal light reflections in the glass face.

---

* It is desirable, of course, to apply signals of amplitude greater than this minimum value; a pulse giving a deflection of ten spot widths will provide a good display for most applications.

6.2.5.1 Effect of Electron Gun Voltages on Spot Size and Deflection Sensitivity

The previous argument has stressed the importance of obtaining a small spot size, and much work has been done to this end on the design of electron guns. The theoretical aspect has been well treated by several authors (see LANGMUIR [607], MALOFF and EPSTEIN [608],

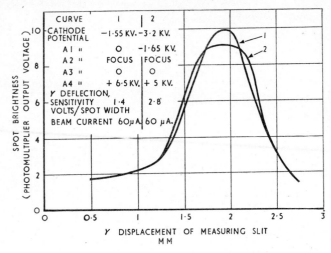

Fig. 6.11. Intensity distribution across spot.

Moss [609]) and only the effect of the gun accelerating voltage on the spot size will be considered here, since it is of direct importance in the design of oscilloscopes for millimicrosecond pulse displays. This is because the accelerating voltage determines the electron velocity through the deflecting plate structure, and therefore affects the frequency bandwidth of the cathode ray tube.

The spot size actually obtained from an electron gun is at variance with that predicted by the conventional electron optical theories, since it is very difficult to take full account of the various lens aberrations and space charge effects in the beam. The electron gun used in a large number of cathode ray tubes is usually regarded as a double lens system, in which the cathode, grid and first anode form an "immersion" lens, whilst the following two anodes form a convex lens system, which focusses the electron beam on to the screen. If, in such a system, the final anode voltage is altered, conventional theory indicates that the spot size should be inversely proportional to the square root of this voltage. Such is not the case in practice and LIEBMANN [610] has shown that the spot size is substantially independent of the final accelerating voltage; fig. 6.12 is a reproduction of his experimental figures. The result agrees with the theory advanced by LIEBMANN, in which he considers this type of gun as comprising a three lens, rather than a two lens, system. The amount of published experimental evidence concerning the relationship of spot size and beam accelerating voltage is, however, rather limited.

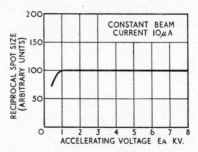

Fig. 6.12. Effect of accelerating voltage on spot width (after LIEBMANN).

Let us now consider an electron beam passing through a pair of deflecting plates, of axial length l. From equation 6.5 we have for the steady state deflection Y at the screen

$$Y = \frac{E}{b} \cdot \frac{e}{m} \cdot \frac{S}{v_x} t_r$$

Now

$$t_r = \frac{l}{v_x}$$

and

$$v_x \propto \sqrt{V_a}$$

where V_a is the gun accelerating voltage, thus

$$Y \propto \frac{1}{V_a}$$

for a given applied signal voltage E.

Hence, as V_a is increased, the deflection is reduced and, since the spot size remains substantially constant for a given value of beam current, the deflection sensitivity, (in terms of deflecting voltage per spot width), is reduced in proportion to the increase in V_a. It should be noted that in this case the frequency bandwidth of the deflecting plates will be increased as V_a increases, due to the reduction of transit-time.

A further corollary of interest is that if the transit-time of the beam through the plates is kept constant, by alteration of the plate length as V_a is altered, then the deflection sensitivity will be proportional to $1/\sqrt{V_a}$. Thus, for a fixed bandwidth,* the deflection sensitivity is inversely proportional to the square root of the accelerating voltage.

6.2.5.2 Effect of Spot Size on the Dimensions of the Display

One factor which influences the deflection sensitivity, and which has not yet been considered, is the spacing between the deflecting plates. As this spacing is reduced the deflection sensitivity is increased so that the distance between the plates should be small (bearing in mind the effect of capacitance on bandwidth). However, if the spacing is reduced too much, the electron beam will impinge on the plates when the beam is deflected so that the minimum plate spacing will be fixed by the maximum required beam deflection. It should be particularly noticed that the maximum deflection is to be defined in terms of spot width, since this will determine the resolution and measurement accuracy. In practice, a maximum deflection of 50 to 100 spot widths in both signal and time base directions will permit a measurement accuracy of 1 or 2 per cent, which is usually sufficient in millimicrosecond pulse applications.

6.2.6 Brightness.

It is axiomatic to say that the brightness of the display must be sufficient to be visible to the eye or else to be recorded by photographic or other means. We shall now briefly consider the various factors which influence the intensity.

6.2.6.1 Beam Power

The primary factor which affects the brilliance is the beam wattage, i.e. the product of beam current and beam accelerating potential. Either or both of these quantities may be increased in

---

* It may be pointed out that changes in the length of the plates will influence their capacitance, which in turn may affect the bandwidth.

order to obtain a brighter spot. An increase in current will, in general, give a larger spot diameter and a correspondingly reduced deflection sensitivity, while, as we have seen, an increase in voltage will also reduce the sensitivity. In tube design, it is usual to employ as high an accelerating potential as is practicable and then arrange that the current supplied by the gun gives sufficient brightness. This procedure is adopted in the case of a tube intended to display single transient pulses, but, when the tube is to delineate recurrent waveforms at a high repetition frequency a greatly reduced spot brightness is adequate and only a low beam current is required. The very small spot width of 0·001 in. was obtained by PIERCE by reducing the current. Such reduction involves the use of beam limiting apertures in the electron gun, and problems of alignment set a minimum value to the spot diameter attainable in practice.

6.2.6.2 Aluminizing

When a high voltage beam of electrons impinges on a screen, a negative charge will be acquired and the potential of the surface

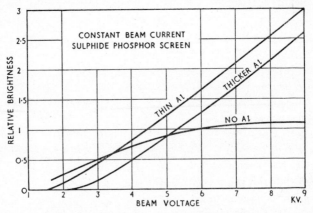

Fig. 6.13. Performance of aluminized screens.

may fall below that of the final accelerating electrode of the electron gun. The beam of electrons will be retarded as the screen is approached with a consequent loss of brilliance. An equilibrium condition is established when the charge reaching the screen, through the primary electron beam, is equal to the loss of charge by the emission of secondary electrons (collected by the last anode of the electron gun), and by the loss due to surface resistance leakage. DUDDING

[611] has discussed this effect, and the results obtained with a sulphide phosphor screen are indicated in fig. 6.13. It will be seen that the effect is not serious until the accelerating voltage rises above 4 KV. This is because fewer secondary electrons can escape from the phosphor due to the greater depth of penetration of the primary electron beam into the phosphor material at the higher voltages. To overcome this "phosphor sticking" the phosphor must possess a low value of transverse resistance, so that the charge collected may leak away between successive time bases strokes. This has been achieved by the use of an aluminium film in contact with the phosphor; the film is sufficiently thin that the electron beam, which transverses the film before striking the phosphor, does not lose much energy. The aluminium film is connected to the final accelerating electrode of the gun, so that the phosphor surface is held at the appropriate potential. The film also improves the brightness by reflecting some of the light from the phosphor into the forward direction towards the observer's eye.

6.2.7 Beam Acceleration After Deflection.

During recent years, it has become common practice to accelerate the electron beam after it has been deflected by the signal plates. Two arrangements have been described by PIERCE [612] and ALLARD [613]. Additional anodes, in the form of annular rings of graphite or other material, are deposited on the glass walls of the tube; these anodes are held at successively higher potentials between the deflecting plates and screen of the tube. In the second method, the deflected beam travels in a field free space until it reaches a grid made of very fine wires. This grid is located close to the screen and is held at the mean potential of the deflecting plates. After passing through the grid, the beam is accelerated rapidly towards the screen by a high voltage applied between it and the screen (which is aluminised). In the latter arrangement the spot size and deflection sensitivity of the tube will be unaffected (ideally) by this additional acceleration, since it occurs over a very small fraction of the path traversed by the beam. It is found in practice, however, that with the annular ring type of post acceleration anode, the deflection is decreased if the accelerating voltage is raised, whilst with the grid method, the deflection is unaltered but the spot size is increased due to the effect of secondary electrons from the grid striking the phosphor. Thus, in both cases, the deflection sensitivity is reduced as the post accelerating voltage is raised to obtain greater spot brightness. The bandwidth will be unaltered because it is determined by the transit-time through the deflecting plates and by connecting lead effects. This result should be compared with that obtained in § 6.2.5.1, where the accelerating voltage was considered to be applied before the electrons reach the deflecting plates. In this case it was found that the deflection sensitivity was inversely proportional to the square root of the voltage, provided the bandwidth of the deflecting plate structure was held constant by alteration of the deflecting plate length.

It follows that, for a given bandwidth, overall accelerating voltage and beam current (i.e., given brightness) there is little difference in deflection sensitivity, whether the voltage be applied before or after deflection. In practice, however, some post acceleration is often used, since it reduces the potentials (with respect to earth) applied to the electron gun, and so eases insulation problems in the tube base. Post acceleration tubes also give a larger display for a given pulse input; the larger display is, of course, more convenient for observation.

6.2.8 Sealed versus Unsealed Cathode Ray Tubes for Photographic Recording.

A considerable number of very fast writing oscilloscopes have been constructed in which the electron beam is allowed to fall directly on to the photographic sensitive material, which is placed inside the cathode ray tube. The tube must be continuously evacuated, since some gas is given off by the photographic materials, and also gas leakage occurs when the film is inserted and removed from the tube. The need for this continuous pumping makes the equipment very bulky and expensive, so that it is preferable, whenever possible, to use the more conventional types of sealed cathode ray tubes. No pumping equipment is required, and since the photographic film is outside the tube, it is easy to change the films or plates. The recording efficiency, however, of an electron beam writing directly on to a photographic emulsion is considerably greater than when a screen phosphor and optical lens systems are interposed between the electron beam and photographic surface; very high writing speeds may therefore be obtained with an unsealed tube. This statement is only true if the electrons have sufficient energy to penetrate the gelatin layer of the film. An accelerating voltage in excess of 30 KV is usually required, values lying between 30 and 100 KV being normally used (e.g. ROGOWSKI et al. [614]). HERCOCK [615] has discussed photographic recording with such tubes and finds that it is possible to operate at much lower voltages if special types of film are employed. A practical comparison of the effective writing speeds of sealed and unsealed tubes may be obtained by comparing performance figures quoted by Smith and by Lee (see Table 6.1, § 6.2.10).

6.2.9 Photographic Technique.

The photographic technique employed for recording cathode ray tube displays has been fully discussed by HERCOCK [615]. Current practice, in the case of millimicrosecond pulse displays, is to use the largest aperture lens available (f.1) and a fast recording 16 mm or 35 mm film. The application of intensifiers during the film processing does not, in general, give any significant improvement in the maximum possible writing speed which can be recorded.

TABLE 6.1

Parameters	Cathode Ray Tube			
	1	2	3	4
Pre-deflection accelerating voltage (KV)	3	50	2	10
Post-deflection accelerating voltage (KV)	5	—	—	25
Total cathode-to-screen voltage (KV)	8	50	2	35
Maximum writing speed for transient recording (spot widths per second)	2×10^9	2×10^{11}	—	10^{11}
Signal deflection plate bandwidth for 3 db loss (Mc/s)	200	8600	1200	2600
Signal deflection plate input impedance (ohms)	—	—	75	50
Signal deflection sensitivity (volts per spot width)	2·8	10	0·027	0·6
Spot size (mm)	0·9	0·01	0·02	0·15

Tube 1. Normal sealed tube (G.E.C. type VCRX357A) with short connection through glass envelope to deflection plates. Aluminized screen with single post acceleration anode.

Tube 2. LEE's unsealed tube without post acceleration. The electron beam is focussed directly on to the recording film.

Tube 3. PIERCE's sealed tube with four pairs of deflection plates connected as a travelling wave deflection system. This tube is only suitable for displaying recurrent waveforms.

Tube 4. SMITH's sealed tube with helical type travelling wave deflection system. Four post acceleration electrodes are used.

6.2.10 Performance of Some Cathode Ray Tubes.

In the preceding Sections we have discussed in detail the various factors affecting cathode ray tube performance. Some of these factors are mutually conflicting and a compromise must be made. In particular, a small spot size is required in order that the deflection sensitivity may be high, but if the size is reduced too much there will not be sufficient brightness to give the necessary writing speed. The Table (p. 195) of data on four particular cathode ray tubes is given in conclusion so that the reader may appreciate what performance may be obtained with existing tube designs.

6.3 CIRCUIT DESIGN FOR TRANSIENT RECORDING OSCILLOSCOPES

The general functions, which the circuits associated with such oscilloscopes must fulfil, may be stated as follows. Initially, the oscilloscope rests in a quiescent state, with no spot visible on the screen of the tube, until the pulse which is to be recorded arrives at the oscilloscope. The input signal then triggers the time-base circuit which switches on the electron beam and applies a sweep waveform to the X-plates of the tube. The pulse is then transmitted to the signal deflection plates through a delay cable and amplifier, if necessary, the delay being sufficient to ensure that the electron beam has reached its normal brightness and the time-base scan its normal velocity before the pulse to be displayed reaches the deflecting plates. The display is recorded on a photographic film which is continuously exposed to the screen of the tube, no camera shutter being used. At the end of the time base stroke the electron beam is cut off and the time-base circuits returned to normal. In some automatic equipments, the oscilloscope is arranged to be inoperative for a short interval of time after a pulse has been recorded; during this paralysis period, the film is moved through one frame by a drive mechanism. When the film has been stepped forward into a new position the oscilloscope is ready to record another pulse so that a succession of transient waveforms may be recorded. The maximum number of pulses per second that can be recorded is limited (a rate of two a second being convenient), but if a high rate of recording is required, special camera techniques may be employed, involving, for example, driving the film past the screen of the cathode ray tube at a constant velocity.

Fig. 6.14 is a block schematic diagram of a typical transient recording oscilloscope. The basic circuits associated with the display unit will first be described, and oscilloscopes suitable for the delineation of recurrent pulses will be considered later.

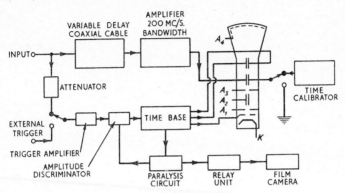

Fig. 6.14. Block schematic diagram of transient recording oscilloscope.

6.3.1 Signal Delay Circuit.

The most convenient form of delay line for millimicrosecond pulse applications is a simple coaxial cable; such cables give delays of the order of 10 mμsec. for 8 ft. The required length may be as much as 200 ft. in some cases, so that it is important to use a cable which gives rise to as little high frequency attentuation as possible. Semi-air-spaced types of cable (embodying a spiral insulator) are normally employed for this purpose; typical attenuation figures are given in § 2.2.6.2.

It will be appreciated that it is important to reduce the amount of signal delay required to a minimum so that the pulse shape to be displayed is not unduly distorted by the cable. The principal factors contributing to the delay are:

(a) that in switching on the electron beam to full brilliance,

(b) the time interval required for the time base waveform applied to the X-deflecting plates to attain its normal velocity of scan,

(c) the interval between the start of the pulse and the instant at which the amplitude has risen to an amount sufficient to trigger the time-base circuit.

Of these delays the first two are controlled by the time-base circuit and should be arranged to run concurrently rather than in series. In some circuit designs the time delay (a) is comparatively long and may be of similar magnitude to the time-base duration; in such a case it is essential to switch on the electron beam before the time-base waveform generator is triggered.

The third delay (c) is a function of the rise time of the input triggering pulse. If the latter has rather a slow rate of rise, then correspondingly large delays will be introduced. The minimum signal delay inserted in the oscilloscope must then be sufficient to cover the maximum time delay that may be introduced by this rise-time effect together with the time-base circuit delays (a) and (b). In practice, a delay time slightly longer than this should be employed, since it is desirable that a small portion of the time-base be visible before the commencement of the pulse waveform. This undisturbed portion of the time base gives a zero reference level for pulse amplitude measurements.

The time delays introduced by time-base circuits at present in use vary between 10 and 200 mμsec., the longer delays obtaining with thyratron circuits.

It may be stressed that a time-base circuit which gives the minimum time delay is the most suitable for use in transient recording oscilloscopes, but this statement does not apply to oscilloscopes employed to display recurrent waveforms.

6.3.2 Time-Base Circuits.

The circuits which produce the pulse for switching on the electron beam and those for generating the time-base saw-tooth waveform will be described together, since they are bound up with one another. Only time-base waveforms of the sawtooth type will be considered, since other forms of time-base find little practical application in the millimicrosecond range of interest.

6.3.2.1 "Bootstrap" circuit

This circuit is an extension of a commonly used linear time-base for slow speed oscilloscopes. It is convenient because it gives a well controlled linear time-base which can be easily switched to give a range of time-base durations, and which introduces only a short time delay. Fig. 6.15 gives the basic configuration of a circuit designed by G. A. HOWELLS [616]; the mode of operation is as follows. The valve V_2 passes about 100 mA of anode current in the quiescent condition, this current being determined by the cathode load resistance and the negative supply voltage. The cathode follower V_4 and the phase inverter V_5 take a current of about 10 to 20 mA. Valves V_1 and V_2 form a cathode coupled flip-flop circuit which is triggered by applying a negative pulse to the anode

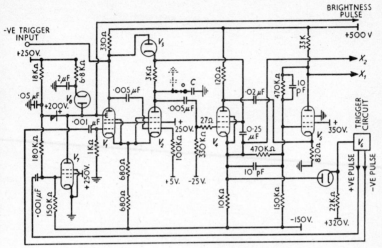

Fig. 6.15. "Bootstrap" time-base circuit.

V_1 EFP60 V_3 CV493 V_5 CV2127 V_7 CV138
V_2 Two valves type CV2127 in parallel V_4 EL37 V_6 EFP60 V_8 CV135

of V_1. This pulse is usually derived from a preceding amplitude discriminator circuit. When the circuit is triggered, the cathode current flowing in V_2 is transferred to V_1, so that a positive pulse is obtained from the dynode of the latter. A current of between 200 and 300 mA is available from this dynode so that a rapidly rising output pulse is obtained. This pulse is applied to the modulation grid of the cathode ray tube to switch on the electron beam; the diode V_8 limits the pulse amplitude to about 60 V. The reader should note the existence of the current multiplication given by V_1 between the cathode (current value of 100 mA) and the dynode (output current of between 200 and 300 mA).

The anode current of V_2, initially of value I_a say, has been cut off when the circuit was triggered so that this current now flows into the capacitor C. The anode potential commences to rise at a rate I_a/C and this rise is passed through the cathode follower V_4 and fed back to the top end of the anode resistor of V_2 so that the voltage across this resistor is kept constant. The current through the resistor is accordingly maintained at the value I_a and a linear rise in voltage is obtained across C. The output from the cathode follower is applied to the X_2 deflecting plate of the cathode ray tube. The valve V_5 is a phase inverter, which is used to drive the X_1 deflecting plate, and so push-pull time base deflection is obtained. Several details are of general interest in the design of millimicrosecond pulse circuits. For example:

(a) During the time base stroke the diode V_3 is cut off and the cathode follower V_4 must supply the charging current $I_a =$ 100 mA for the capacitor C. The current demanded from the cathode follower will be equal to this current plus that required to charge the stray capacitance to earth of the X_2 deflecting plate and associated circuit connections. The total may be as high as 300 mA, with fast time bases, so that there will be a non-linear portion at the commencement of the sweep, whilst the cathode follower is being driven from its quiescent condition (20 mA current) up to the operating current of 300 mA. If the time-base is switched to a low speed range by increasing the anode capacitor C to a larger value, then this non-linear portion will persist since the cathode follower must continue to supply the charging current. The load on the cathode follower, and therefore the duration of the initial non-linearity, may be reduced in such a case by transferring the X_2 plate lead from the cathode of V_4 to the anode of V_2 and deducting the capacitance of the deflector system from the new value of C. This procedure cannot be adopted in the fastest circuits* since the X_2 plate lead will add too much capacitance to the anode of V_2.

(b) The phase inverter valve is driven into heavy anode current during the time base sweep and the current required, 200–300 mA, with the fastest time-base, will be that necessary to discharge the anode and X_1 deflecting plate capacitances at the required rate. A non-linearity will again be introduced at the start of the waveform due to the time taken to bring the

---

* C will be small in value and may even be comprised solely of the stray capacitances at the anode of V_2.

valve from its initial quiescent current of 20 mA up to the required operating current. To reduce this effect the valve may be pulsed quickly into current at the start of the time-base, and this has been achieved in the circuit of fig. 6.15 by deriving a pulse from the anode of V_4 and feeding it into the cathode of V_5.

(c) The non-linear portion at the start of the time-base will occur whilst the electron beam is being switched on; it is found, in practice, that the time-base has become nearly linear before the cathode ray tube spot has attained its normal brilliance.

Up to this point we have only considered the operation of the circuit just after the start of the time-base. Once the circuit has been triggered, the cathode of V_4 will continue to rise at the rate corresponding to the desired time-base velocity, until the cathode to anode potential of this valve falls to a low value such that the valve can no longer pass the required current. The cathode will then cease to rise, and the time-base voltage will remain approximately constant until the cathode coupled trigger circuit V_1, V_2 returns to its quiescent condition. If this saturation effect is allowed to occur, a very bright spot or short line will be present at the end of the time base. This is always undesirable and valves V_6 and V_7 have been introduced to overcome the effect. The valve V_6 is a trigger circuit, as described in § 4.3.1, which operates when the time-base voltage from the cathode follower V_4 reaches a given value, less than that giving rise to saturation. The output from V_6 is used to return the valves V_1, V_2 to their quiescent state, thus resetting the whole circuit. The valve V_7 is pulsed into current, giving a fast back edge to the cathode ray tube modulator pulse derived from the dynode of V_1, and thus switching off the beam.

This circuit was designed for use with cathode ray tube number 1 of Table 6.1 and, in this application, gave a push-pull time-base of 500 V total amplitude with a minimum duration of 0·1 μsec.

A duration of 0·05 μsec. can be achieved by careful reduction of wiring capacitances in the circuit and by the removal of switches for the control of time-base durations.

The starting delay of the circuit, i.e., the time interval between the leading edge of a fast rising trigger input pulse and the cathode ray tube spot attaining its normal velocity and brightness, amounted to about 10 mμsec.

A disadvantage of this particular circuit is the large value of the current taken in the quiescent condition. This limits the application of the circuit to time-base durations of 0·1 μsec. and longer, when used with cathode ray tubes requiring several hundred volts of time base deflection. Similar circuits have also been used by KELLEY [617], and by MOODY and McLUSKY [618].

6.3.2.2 Pulsed Valve Circuits

A different arrangement has been used by BAUER and NETHERCOT [619] which is capable of giving much faster time-base sweeps

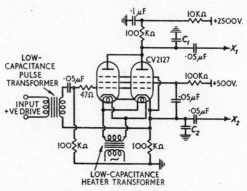

Fig. 6.16. Driven valve time-base circuit.

than is the circuit of fig. 6.15. The time-base velocity, however, is not accurately controlled so that a time calibration waveform must always be included if accurate measurements are required. The method consists of pulsing a valve into heavy current and allowing this current to flow into the time-base capacitor; in the typical circuit, depicted in fig. 6.16, a pulse is applied between the grid and the cathode of a valve via a low capacitance pulse transformer. The duration of the driving pulse determines the duration of the sweep, and the same pulse is often used to brighten up the cathode ray tube. The time-base waveform may be taken from both the anode and the cathode to obtain a push-pull output, the time-base velocity being determined by the magnitude of the anode and cathode capacitors C_1 and C_2 and by the valve current. The capacitors may be chosen so that the anode and cathode waveforms are equal in amplitude. For the fastest time-bases these capacitors degenerate into stray capacitances.

A similar driven valve circuit has been used by SMITH [620]

(fig. 6.17) for providing fast time-bases in LEE's micro-oscillograph. In this application a 3D21 type of pentode was employed to give a time-base waveform at its anode, and the valve was driven into saturation current by a 2D21 type thyratron. An anode current of up to 10 A was obtained and was controlled by varying the heater voltage. By this means the time-base duration for a 1000 V sweep could be varied from 4000 mμsec. (at 2·8 V heater) to a minimum value of 4 mμsec. (at 6·3 V heater supply). For heater voltages giving durations of from 10 to 30 mμsec. the time-base velocities

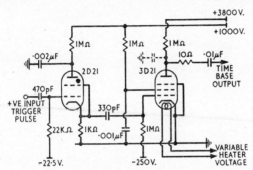

Fig. 6.17. Saturated valve time-base circuit.

were found to be somewhat non-linear, but excellent linearity was obtained outside this range. The 3D21 valve produced a total time-base amplitude in excess of 2000 V; the first 1000 V of this sweep was not used.

6.3.2.3 Thyratron Circuits

Hot cathode thyratron time-base circuits are not very suitable for use in transient recording oscilloscopes because of the comparatively long firing delay which exists between the application of a trigger pulse and the occurrence of the discharge between the anode and the cathode of the valve. However, it is found in practice that this time delay (of the order of 0·1 μsec.) is remarkably constant in value, provided the trigger pulse has a large amplitude with a fast rising leading edge. The delay is dependent on the power input to the control grid and the trigger pulse should drive at least 20 mA of grid to cathode current through the valve in the case of a small thyratron, and correspondingly more for a larger valve.* Under

* If no grid current is taken, the input pulse being just sufficient to trigger the thyratron, then variable delay times as long as 1 μsec. will be obtained.

these conditions the thyratron can be used as a time-base generator, and such circuits find application in oscilloscopes for the display of waveforms which are regularly preceded in time by a pulse, suitable for triggering purposes. This subject will be discussed in § 6.4.

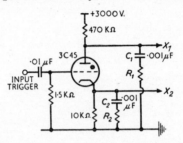

Fig. 6.18. Thyratron time-base circuit.

The basic circuit of a thyratron time-base is shown in fig. 6.18 and PRIME and RAVENHILL [621] and others have described typical applications.

Some little time after the thyratron has been triggered,* the anode to cathode current increases rapidly to a maximum, in a time of

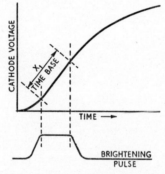

Fig. 6.19. Thyratron time-base waveforms.

about 10 mμsec. The current flows through the capacitors C_1, C_2, the resistors R_1, R_2 being used to limit the amplitude of the current to a safe value. Several amperes may be obtained from the hydrogen thyratron circuit shown in the figure. The anode and cathode waveforms are used to drive the X-deflection plates of the cathode ray tube, the capacitor sizes being chosen to give the required time-base speed. The time-base waveform (fig. 6.19) will be exponential

---

* The small thyratron type 2D21 is suitable for use in the triggering circuit.

in shape since the current through the condensers diminishes as the time-base progresses; in addition there is the slow start alluded to above due to the time taken by the thyratron to become fully ionized. The portion of this waveform lying between the end of the starting period and the point corresponding to about one third of the peak amplitude is approximately linear, and is used as the display time-base. This segment of the waveform may be selected by applying a suitable pulse to the modulator grid of the cathode ray tube such that the beam is only switched on during this portion of the total time-base waveform. If used in this manner, the total voltage generated by the circuit must be arranged to be at least three times as great as that required for the cathode ray tube time-base sweep.

The circuits for providing the brightening waveform for the cathode ray tube employ normal types of pulse generator and will not be described. It is stressed that this pulse must be phased correctly with respect to the time-base waveform.

The principal advantage of a thyratron circuit is that a current of several amperes is available for the time-base, so that very fast sweep speeds may be obtained; the current is also well controlled by the resistors present. The time-base waveform, however, is non-linear and must be carefully calibrated before it can be used for accurate time measurements.

6.3.3 Auxiliary Circuits.

Various additional circuits, indicated on the block schematic diagram of fig. 6.14 are found to be necessary. The need for this auxiliary apparatus will depend on the particular application, but it is desirable that these circuits should be readily available and, if possible, built into the oscilloscope. This equipment comprises the following:

6.3.3.1 Signal and Trigger Amplifiers

The design of wideband amplifiers has already been treated in Chapter 5. An amplifier would, of course, only be employed if the amplitude of the waveform to be displayed is too small to be applied directly to the signal deflection plates, but, if the waveform is also to trigger the time-base, then an additional amplifier may be needed to obtain a pulse sufficiently large for this purpose. The bandwidth of this auxiliary amplifier can generally be restricted to less than 50 Mc/s.

In cases where the time-base circuits must be triggered by the pulse that is itself to be displayed, care is necessary to ensure that the connection to the auxiliary circuits from the input terminal does not affect the shape of the signal waveform. This requirement will often necessitate the use of an attenuator (indicated in fig. 6.14) in the cable connection to the time-base circuits. The attenuator, introducing 10 to 20 db loss, will minimize the effects of loading, due to small stray capacitances, and of reflected pulses from slightly mis-terminated coaxial cables. The attenuation must be offset by the provision of additional gain in the trigger amplifier.

6.3.3.2 Amplitude Discriminator

In some applications it is desirable that only input pulses of amplitude greater than a certain amount should trigger the time-base circuit. This may be achieved by the use of an amplitude discriminator circuit preceding the time-base unit and such circuits are described in Chapter 7. The reader should notice that the provision of such a discriminator, in combination with the signal amplifier and trigger amplifier gain settings, affords a means whereby the time-base is only triggered by pulses the amplitudes of which are large enough to give a satisfactory display on the cathode ray tube. This facility is useful in applications where successive input pulses vary in amplitude between wide limits.

6.3.3.3 Time Calibration Circuit

The fast time-bases used often suffer from non-linearity, particularly when durations of less than $0 \cdot 1$ μsec. are concerned, so that it is essential to calibrate the time-base if accurate time measurements are required. This calibration may be simply performed by recording a single trace showing either a sine wave derived from a signal generator or a series of short pulses. The frequency of this sine wave or the recurrence frequency of the pulses is adjusted to suit the time-base duration and is measured accurately.

Fig. 6.20 shows a simple circuit (WELLS [622]) which has been used to generate pulses for a time calibration waveform. The secondary emission valve functions as an oscillator by inserting a tuned circuit in the dynode lead and applying positive feed-back from dynode to control grid; frequencies of up to 100 Mc/s may be generated. The output is taken from the anode across a low value resistor and consists of short duration pulses at a recurrence frequency determined by the tuned circuit. A sine wave of small amplitude is superimposed on the pulse output, due to the valve capacitance between anode and dynode, but this may be removed by means of a biassed germanium diode.

At frequencies above 100 Mc/s a sine wave is generally employed for time calibration purposes.

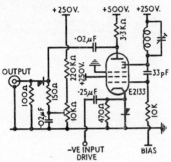

Fig. 6.20. Time calibration pulse generator.
(From *Nucleonics*, Volume 10, No. 4, pages 28-30.
Copyright 1952, McGraw-Hill Publishing Company Inc.)

6.3.4 Camera Equipment.

A standard type of camera, with shutter mechanism, may be used for photographic recording with oscilloscopes displaying recurrent waveforms. When recording transients however, no shutter is employed and the film is continuously exposed to the screen of the cathode ray tube; means must be provided for moving the film, either frame by frame, or at uniform speed. When the time interval between input pulses is long, of the order of one minute, the usual practice is to move the film forward one frame after each time-base stroke. In either case, a paralysis circuit must be inserted in the system, so that the time-base circuit is inoperative for a sufficient period after each single stroke in order to allow the film to be moved into a new position. This is necessary to avoid the confused picture recorded when two time-bases are superimposed. The paralysis circuit usually forms part of the amplitude discriminator; a suitable arrangement is detailed in § 7.5.

6.4 OSCILLOSCOPES FOR THE DISPLAY OF RECURRENT WAVEFORMS

In the preceding Sections we have been concerned with the design of time-base circuits and cathode ray tubes for the display and photographic recording of irregularly occurring waveforms. The design differs from that required when recurrent pulses are to be delineated in the following main respects:

(a) The brightness of a single trace must be sufficient to give a photographic recording.

(b) The time-base starting time delay must be a minimum.

(c) Different photographic recording techniques are adopted in the two cases.

The first two requirements need not be met where recurrent pulses are concerned. The brightness of the cathode ray tube trace will depend on the recurrence frequency of the pulses to be displayed and the tube may be operated at correspondingly reduced beam current and spot size. A tube designed specifically for this application is number 3 of Table 6.1, page 195; the increase in deflection sensitivity obtained should be noted.

The time-base starting time delay is not significant* when regular pulses are to be displayed, since auxiliary delay circuits can generally be included to provide the required time-base trigger pulses at any suitable time (of the order of 1 μsec.) preceding the waveform to be observed. By suitable adjustment of the delay circuits the waveform to be measured can be adjusted to occupy a central position on the display. The time-base circuits previously described may be used and the thyratron type of circuit (§ 6.3.2.3) is suitable for this application, when very short duration time-bases are required.

An alternative and elegant method of displaying recurrent waveforms is described in the next Section.

6.5 OSCILLOSCOPES FOR RECURRENT PULSES USING PULSE SAMPLING TECHNIQUES

Such types of oscilloscope differ fundamentally from those previously discussed and their basic principle of operation (JANSSEN [623]) is related to that of a stroboscope used for viewing high frequency phenomena. McQUEEN [624] has developed an instrument possessing a signal deflection bandwidth in excess of 300 Mc/s; performance is better than that obtained with simple types of cathode ray tube, but is not as good as that achieved by the travelling wave deflector

---

* The delay may be large but must be free from random variations in magnitude.

plate design described in § 6.2.4.3. However, this bandwidth performance is obtained without using specially designed cathode ray tubes and with a minimum of high frequency circuits. The sensitivity obtained is high so that there are great practical advantages in the method and future developments may well give greater deflection bandwidths. The instrument cannot be used for displaying irregular pulses.

The technique of pulse sampling may be illustrated by reference to fig. 6.21. In this diagram, the second line shows a typical short

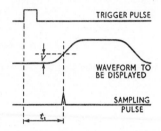

Fig. 6.21. Principle of operation of pulse sampling technique.

duration pulse waveform which it is required to display and the third depicts a sampling pulse occurring at time t_1 after a reference time-zero derived from the input trigger pulse. The amplitude of the waveform under test is measured during the sampling pulse interval by a suitable mixing circuit, which gives an output proportional to the average amplitude of the test waveform throughout the duration of the sampling pulse. This output is used to control the amplitude of a long pulse, the duration of which is made a little less than the recurrence period of the test waveform. The long duration pulse then provides the Y deflection on the display cathode ray tube, the X deflection being made proportional to the time interval t_1 (see fig. 6.22). If now t_1 is altered between successive pulses then the amplitude at different points along the test waveform will be measured and the Y deflection on the display tube will vary in accordance with the test waveform. In practice, t_1 is increased in a linear manner from a minimum value to a maximum, and is then quickly restored to the minimum value again. This saw-tooth change of t_1 is then repeated ad infinitem. The recurrence frequency of the display is equal to that of the saw-tooth. The time taken for the delay t_1 to change from its minimum value to its maximum, divided by the recurrence period of the test waveform, determines the number

of elements of the test waveform measured and hence defines the
number of pulses building up the display on the cathode ray tube.
Fig. 6.22 depicts a display containing only 10 elements for clarity
in drawing but in practice this number should not be less than 100.
Also, in a practical oscilloscope, only the peaks of the pulses applied
to the Y-plates of the tube are made visible by switching the electron
beam on and off at suitable instants.

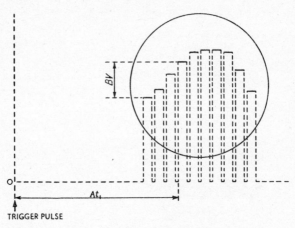

Fig. 6.22. Display principle of pulse sampling oscilloscope
(A and B are proportionality constants).

The final picture on the screen appears as a waveform made up of a
series of short horizontal lines with small gaps between adjacent
"dashes".

The shortest duration pulse or the highest frequency sine waveform which this type of oscilloscope can display, without undue
distortion, is entirely controlled by the high frequency performance
of the initial mixing circuit and the duration of the sampling pulse.

6.5.1 Mixing Circuits.

A variety of possible circuits may be employed in the mixing
stage. The circuit used by McQueen, shown in fig. 6.23, is satisfactory at frequencies up to at least 300 Mc/s. When much higher
frequencies are to be handled, diode mixing circuits or valves
suitable for use at 1000 Mc/s must be employed in order to avoid
frequency limitations due to transit-time and connecting lead inductance. Referring to fig. 6.23, the valve is normally biassed off and is

pulsed into current by the negative sampling pulse applied to the cathode. The bias and pulse amplitude are adjusted so that, on the peak of the sample pulse, an anode current of about 8 mA is passed (the normal value when the valve is used as an amplifier). The exact value of the current will be controlled by the grid potential which is derived from the waveform under test via a capacitance potentiometer. The attenuation is adjusted such that the test waveform does not exceed 0·5 V peak to peak amplitude at the grid of the

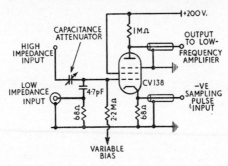

Fig. 6.23. Mixer circuit for pulse sampling system.

mixing valve. The anode current thus consists of a series of pulses, modulated in amplitude by the test waveform, the maximum amount of variation being kept small by restricting the test waveform amplitude applied to the grid of the valve. This restriction is necessary to ensure that the valve characteristic is approximately linear over the working range and so that the changes in anode current will be a linear function of the amplitude of the test waveform. The current is integrated by capacitance at the anode and the voltage output from the mixing valve is then amplified by a low frequency amplifier. A comparatively long duration pulse is thus obtained, with an amplitude proportional to the mean value of the pulse of anode current flowing in the mixing valve.

In this particular mixing circuit it is necessary to provide a capacitance (including stray capacitance), from grid to earth to offset the effect of that between the grid and the cathode. The latter valve capacitance and the grid to earth capacitance form a potential divider across the cathode input drive, so that the amplitude of the pulse appearing between grid and the cathode of the valve will be less than that supplied by the sampling pulse generator.

This attenuation effect must not be too great, so the total grid to earth capacitance should not be less than 5 pF.

6.5.2 Frequency Limitation depending upon Sampling Pulse Duration.

In practice, the shape of the sampling pulse will probably approximate to a half sine wave before application to the mixing valve; it will become distorted, however, by non-linearity in the mixing valve near cut-off, so that we may assume the pulse of anode current to be triangular. If H is the voltage amplitude (measured above cut-off) and τ is the duration of the sampling pulse, and if $V(t)$ is the amplitude of the grid signal to be displayed, we have for the anode current at any instant t

$$I_a(t) = 0 \quad \text{for} \quad t \leqslant t_1 - \tau/2$$

$$I_a(t) = g_m \left[V(t) + \frac{2H}{\tau}(t - t_1 + \tau/2) \right] \text{ for } t_1 - \tau/2 \leqslant t \leqslant t_1$$

$$I_a(t) = g_m \left[V(t) + \frac{2H}{\tau}(t_1 + \tau/2 - t) \right] \text{ for } t_1 \leqslant t \leqslant t_1 + \tau/2$$

$$I_a(t) = 0 \quad \text{for} \quad t_1 + \tau/2 \leqslant t \qquad (6.21)$$

We have here supposed that the valve, of mutual conductance g_m, behaves linearly, once it has been driven into the conducting state by the triangular sampling pulse.* The capacitance C at the anode integrates this current and a voltage fall V_a is produced at the end of the sampling pulse, given by

$$V_a(t_1) = \frac{g_m}{C} \int_{t_1-\tau/2}^{t_1+\tau/2} V(t)dt + \frac{2Hg_m}{C\tau} \left[\int_{t_1-\tau/2}^{t_1} (t - t_1 + \tau/2)dt \right.$$
$$\left. + \int_{t_1}^{t_1+\tau/2} (t_1 + \tau/2 - t)dt \right] \quad (6.22)$$

The term in brackets represents an output pulse of constant amplitude, independent of the input signal, and we are only concerned with the added contribution due to the first term.

---

* Strictly, the width of the sampling pulse is not constant but varies with the amplitude of the signal under test; this is because the instants of time at which the valve starts to conduct or is cut off are determined by the instantaneous difference between the potentials of the grid and cathode.

6.5.2.1 Pulse Conditions

Consider the application at time $t = t_0$ of a voltage step function input of amplitude V (fig. 6.24). The first term in equation 6.22 then gives:

$$V_a(t_1) = 0 \quad \text{for} \quad t_1 + \tau/2 \leqslant t_0$$

$$V_a(t_1) = \frac{g_m}{C} \int_{t_0}^{t_1+\tau/2} V dt = \frac{V g_m}{C} (t_1 + \tau/2 - t_0)$$

$$\text{for} \quad t_1 - \tau/2 \leqslant t_0 \leqslant t_1 + \tau/2$$

$$V_a(t_1) = \frac{g_m}{C} \int_{t_1-\tau/2}^{t_1+\tau/2} V dt = \frac{V g_m \tau}{C} \quad \text{for} \quad t_0 \leqslant t_1 - \tau/2 \quad (6.23)$$

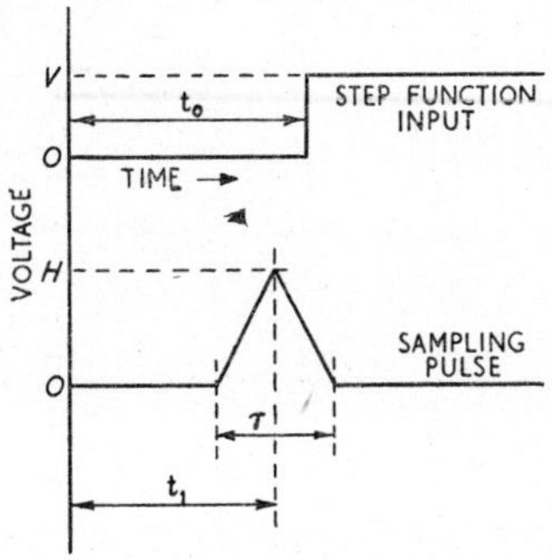

Fig. 6.24. Sampling pulse applied to step function input.

The deflection $Y(t_1)$ on the display tube, which is proportional to $V_a(t_1)$, is accordingly given by

$$\frac{Y}{Y_0} = 0 \qquad \text{for} \quad t_1 + \tau/2 \leqslant t_0$$

$$\frac{Y}{Y_0} = \frac{t_1 - t_0}{\tau} + \frac{1}{2} \quad \text{for} \quad t_1 - \tau/2 \leqslant t_0 \leqslant t_1 + \tau/2$$

$$\frac{Y}{Y_0} = 1 \qquad \text{for} \quad t_0 \leqslant t_1 - \tau/2 \quad (6.24)$$

where $Y_0 = V g_m \tau / C$ is the steady response, neglecting the additive constant terms in equation 6.22 already alluded to.

The response curve is depicted in fig. 6.25, and it is seen that the rise-time of the displayed pulse cannot be greater than the sampling pulse width τ.

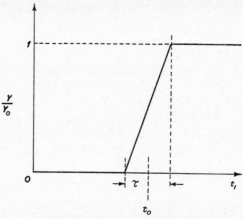

Fig. 6.25. Theoretical response of pulse sampling oscilloscope to step function input at time t_0.

6.5.2.2 *Sine Wave Conditions*

Let us now consider the case when the input signal is in the form of a sine wave

$$V = V_0 \sin \omega t \quad \text{(see fig. 6.26)}$$

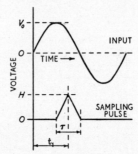

Fig. 6.26. Sampling pulse applied to sinusoidal input.

Relation 6.22 then yields

$$V_a(t_1) = \frac{V_0 g_m}{C} \int_{t_1-\tau/2}^{t_1+\tau/2} \sin \omega t \, dt = -\frac{2V_0 g_m}{\omega C} \sin \frac{\omega \tau}{2} \sin \omega t_1 \quad (6.25)$$

Thus, as t_1 is varied to move the sampling pulse along the sine wave, the deflection on the display will also be sinusoidal in shape. The amplitude $Y_0(0)$ of the displayed oscillation at low frequencies is given by $Y_0(0) = V_0 g_m \tau/C$, whence at a frequency ω

$$\frac{Y_0(\omega)}{Y_0(0)} = \frac{\sin \omega\tau/2}{\omega\tau/2} \tag{6.26}$$

This relation has been plotted in fig. 6.3; it is seen how the amplitude of the display decreases as the frequency is raised.

6.5.3 Display Circuits.

One of several methods of display may be adopted, but, as the circuits involved (except for the sampling pulse generator) use

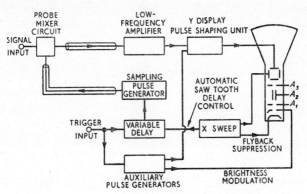

Fig. 6.27. Block schematic diagram of pulse sampling oscilloscope.

standard low frequency pulse techniques, they will not be described in detail. The general schematic diagram is given in fig. 6.27 to show what circuits are required in order to utilize the pulse information available from the mixing unit.

The input trigger pulse triggers the sampling pulse generator through a variable delay circuit which defines the delay time t_1 of fig. 6.21. The slow speed X time-base generator varies this delay time in a saw-tooth manner such that the rate of change of delay t_1 is proportional to the rate of change of the X deflection voltage applied to the display tube. The total change in delay time, from minimum to maximum, will determine the time interval over which the test waveform is sampled and so defines the effective time-base duration of the oscilloscope. Values of from 0·05 to 5 μsec. have been used by McQueen. The duration of the X-sweep on the cathode ray tube determines the number of elements forming the display and a value of not less than one hundred times the trigger pulse recurrence period should be employed. In a practical instrument both the X time-base duration and the range of variation of the delay time t_1 should be controllable at will.

The sampling pulse is fed into the mixing unit through a matched coaxial cable, the mixing circuit being constructed in the form of a small probe unit which may be mounted very close to the source of the waveform under test. The output from the mixer is fed to the main oscilloscope via a low capacitance coaxial cable and is then amplified by a low frequency amplifier. It should be noted that a D.C. amplifier is not required, since the information is still in pulse form. It is necessary to ensure that the amplifier response is linear over the range of "modulation" of the pulses;* some of the valves in the later stages will require to be operated in a class C manner in order to avoid overloading.

The output pulse from the amplifier is then formed into one of the same peak amplitude but of long duration. This long pulse provides the Y deflection on the cathode ray tube. Other suitable waveforms, derived from auxiliary pulse circuits, are employed for electron beam switching including the extinguishing of the spot during the time-base flyback.

One method of operating this type of oscilloscope has been described. Various refinements are possible; two independent waveforms, for example, may be displayed simultaneously on the same cathode ray tube screen. This may be performed by employing two mixing units and switching their outputs in turn into the same low frequency amplifier. Alternate pulses fed into the amplifier will thus be derived from the two different input waveforms and the final display will appear as two independent waveforms, each composed of half the normal number of elements. Alternatively, each mixing unit may be switched into and out of circuit on alternate time-base sweeps; the recurrence frequency of the final display will then be halved. This technique may be extended to three or more mixing circuits controlled by an appropriate switching sequence.

---

* The maximum depth of modulation is not more than 25% of the average pulse amplitude.

6.5.4 Brightness of the Display.

It has been pointed out in the preceding discussion that the recurrence frequency of the display should not be greater than one hundredth of the test waveform repetition frequency. Thus, if the display is to be free from flicker, the input pulse recurrence frequency must not be less than 5000 per sec. If the waveform to be tested has a low recurrence frequency, a cathode ray tube with a long afterglow characteristic may be employed, or alternatively, the display may be recorded photographically.

6.5.5 General.

The input impedance of the mixing probe is determined by the input capacitance potentiometer. In the most sensitive condition, 0·5 V signal gives full deflection on the tube and the input capacitance amounts to about 5 pF. When a lower sensitivity can be tolerated, the capacitance may be reduced to 1 or 2 pF, thus giving rise to very little loading on the source under test—a valuable feature of this technique. In addition, the only circuits handling high frequency signals are located in the mixing valve probe and all high frequency cable connections to the oscilloscope proper are eliminated (except that conveying the sampling pulse). Many possible causes of distortion, encountered in a more conventional oscilloscope, due, for example, to the use of distributed amplifiers and to reflections in imperfectly terminated cables, thus disappear. The localization of the high frequency circuits to the mixing valve probe is one of the major advantages of this type of oscilloscope.

7

APPLICATIONS TO NUCLEAR PHYSICS

7.1 INTRODUCTION

THE basic circuit techniques which are required, when pulses of millimicrosecond durations are to be handled, have been described in detail in the preceding chapters. We shall now show how these, and other more specialized circuits, have been applied to some of the measurement problems encountered in nuclear physics research. The use of the particular circuits to be considered is not necessarily limited to this field, and, although designed, at the present time, primarily for this purpose, such circuits may well find a wider range of application in the future. In the next chapter some further applications will be indicated.

7.2 GENERAL MEASUREMENT PROBLEM

The general measurement problem in nuclear physics, to our knowledge of which the application of millimicrosecond pulse techniques has made a substantial contribution, may be briefly stated as follows:

(a) The measurement of the time interval between the emission of two related nuclear particles, or radiations, from a radioactive nucleus.

(b) The selection of a particular nuclear particle or quantum of radiation and the measurement of its energy and direction of motion.

(c) The measurement of velocity by observing the time taken for a particular particle or quantum to travel between two fixed points.

(d) The measurement of the number of nuclear particles, or quanta of radiation, arriving at a particular element of volume during a given time interval.

The first piece of apparatus needed for these measurements is a "counter" which delivers a voltage or current pulse as soon as possible after the arrival of the particle to be detected. If energy measurements are to be made then the amplitude of the pulse must be related to the energy of the nuclear event initiating it. Having once obtained the pulse, the measurement problem becomes a purely electronic one involving the display and examination of the pulse waveforms and the measurement of (i) their amplitude, (ii) the time interval between related pulses, and (iii) the number of pulses occurring in a given time. This latter measurement is termed "scaling"; the term "counting" will be reserved for the process of detecting a particle or quantum by the formation of a voltage or current pulse.

Two types of counter have been employed to drive fast pulse measuring circuits and these will be described first. The measurements (i), (ii) and (iii) alluded to above will be considered in subsequent sections.

7.3 SCINTILLATION COUNTERS

Counters of this form have been described by many authors and much work has been done on their development. A review article by COLTMAN [701], and the companion volume by J. B. BIRKS in the present series entitled *Scintillation Counters*, may be consulted.

The nuclear particles or quanta are allowed to enter a suitable organic crystal or other phosphor. Some of the atoms are excited and on returning to their normal state emit light (scintillate). There exists a probability distribution for the small time interval which elapses between the initial exciting event and the emission of a light photon from an excited atom. If the rate of photon emission, after the passage of a nuclear particle, is measured it is found to decay to zero; an approximately exponential law is followed, the time-constant of which is dependent on the particular phosphor material employed.

The light from the phosphor is picked up by the photo-cathode of an electron multiplier tube (fig. 7.1) and the resulting photoelectrons are multiplied in the succeeding electron multiplication stages. A burst of electrons arrives at the final collecting anode for each light scintillation of the phosphor, the rate of electron arrival, or current, being controlled by the rate of emission of photons from the phosphor. A current pulse is thus obtained which is related in time to the nuclear event under observation. The time integral of this current pulse, i.e. the charge, is proportional to the total number of photons emitted by the phosphor, which in turn is a function of the energy of the particle or quantum of radiation being measured. The exact shape of the current pulse is of interest since the latter will be used to

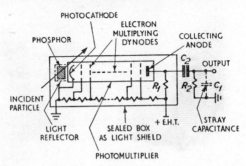

Fig. 7.1. Schematic diagram of scintillation counter.

actuate all the succeeding circuits; factors affecting the rise-time and decay-time will now be considered.

7.3.1 Current Pulse Shape.

Let us take the case of a single photoelectron leaving the photocathode and forming, by the process of electron multiplication, an avalanche of electrons at the collecting anode of the photomultiplier tube. These electrons arrive after a certain time interval following the emission of the initial photoelectron. Furthermore, they do not all arrive at precisely the same instant, but are spread out over a further time interval, because the paths traversed by the individual electrons between multiplying stages are not identical, and so different electrons have different transit-times between stages. The latter time interval alluded to is called the "spread of transit-time", whilst the time which elapses between the start of the photoelectron from the photocathode and the arrival at the collecting anode of half the total number of electrons is simply termed the "transit-time". The rate of arrival of electrons at the anode will vary with time accordingly to a probability law, determined by the geometry of the photomultiplier. We shall assume this law to be Gaussian, giving a good approximation to the current found in practice, thus

$$I = \frac{N_0 e}{\sqrt{\pi} T_m} \cdot e^{-\left(\frac{t-\tau}{T_m}\right)^2} \tag{7.1}$$

where I = current at time t,
N_0 = total number of electrons arriving at the anode per single photoelectron,
e = electronic charge,
τ = transit time,
T_m = constant, determining the spread of transit time.

The constant factor in front of the exponential is chosen such that the following relation is satisfied:

$$\int_{-\infty}^{\infty} I \, dt = N_0 e \tag{7.2}$$

The curve in fig. 7.2 (for the case of zero phosphor decay-time constant T_p) shows the shape of the current distribution; this also gives the average shape of a "noise" pulse due to a single photoelectron leaving the photocathode. In practice T_m usually lies in the range 0·1–10 mμsec. whilst N_0 has a value between 10^5 and 10^8 per single photoelectron.

When the effect of the phosphor decay-time is taken into account the photoelectric emission from the photo-cathode, when a particle is incident at time $t = 0$, will follow a law of the form

$$n = \frac{N}{T_p} e^{-t/T_p} \tag{7.3}$$

where n = number of photoelectrons leaving the photocathode per second,
and N = total number of photoelectrons released by the light scintillation.

This relation is not strictly correct; with some phosphors a sum of two exponential terms, of widely differing time-constants, is a closer approximation. In addition, the photocathode current is not a continuous but a statistical function, as discussed by Post and Schiff [702]. Equation 7.3 is, however, sufficiently accurate for our present purpose.

The anode current I at time t due to a particle or quantum striking the phosphor at $t = 0$ may be obtained by combining relations 7.1 and 7.3:

$$I = \int_0^t \frac{N_0 e}{\sqrt{\pi} T_m} \cdot e^{-\left(\frac{t-t'-\tau}{T_m}\right)^2} \cdot \frac{N}{T_p} e^{-t'/T_p} \, dt' \tag{7.4}$$

This may be written in the form

$$I = \frac{NN_0 e}{\sqrt{\pi T_m T_p}} \cdot e^{[-(t-\tau)/T_p + T_m^2/4T_p^2]} \cdot \int_0^t e^{-\left[t'/T_m - \left(\frac{t-\tau}{T_m} - \frac{T_m}{2T_p}\right)\right]^2} \cdot dt'$$

(7.5)

whence

$$I = \frac{NN_0 e}{2T_p} \cdot e^{[-(t-\tau)/T_p + T_m^2/4T_p^2]} \left\{ \operatorname{erf}\left[\frac{t-\tau}{T_m} - \frac{T_m}{2T_p}\right] + \operatorname{erf}\left[\frac{\tau}{T_m} + \frac{T_m}{2T_p}\right] \right\}$$ (7.6)

Current waveforms are shown in figs. 7.2 and 7.3 for different values of the ratio T_m/T_p—in these curves it has been assumed that

$$\operatorname{erf}\left[\frac{\tau}{T_m} + \frac{T_m}{2T_p}\right] \simeq 1$$

Fig. 7.2 illustrates the effect of altering the phosphor time-constant T_p with a fixed spread of transit-time in the photomultiplier;

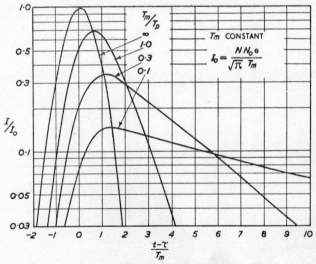

Fig. 7.2. Photomultiplier current output waveform.
(spread of transit-time T_m constant).

fig. 7.3 depicts the improvement in pulse shape obtained when the photomultiplier constant T_m is decreased with a fixed value of T_p. In both these figures the total number of electrons emitted from the photo-cathode, due to the light scintillation from the phosphor, has been assumed to be constant. In practice, T_p can only be varied by a change of phosphor material, and this in turn will give a variation in the total number of photoelectrons available for multiplication.

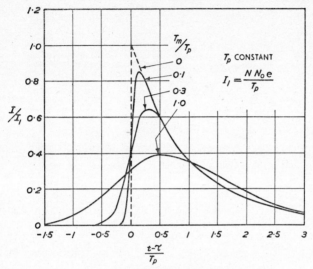

Fig. 7.3. Photomultiplier current output waveform (phosphor decay-time T_p constant).

7.3.2 Performance of Existing Scintillation Counters.

The shape and magnitude of the output current pulse obtained from scintillation counters has been studied by several authors (see POST and SCHIFF [702], COLLINS [703], POST and SHIREN [704], LUNDBY [705], BITTMAN et al. [706], ALLEN and ENGELDER [707]). Such counters are being developed to give shorter pulse durations and the following tables will give the reader an appreciation of the performance of some phosphors and photomultiplier tubes at present in use:

TABLE 7.1. Phosphor decay time-constants

Phosphor	Decay time constant (T_p) (mμsec.)
Naphthalene	87
Anthracene	23
Trans-stilbene	6
p-terphenyl	4·2

TABLE 7.2. Photomultiplier constants

Type	Test voltage (normal operating voltage 1500–2000 V)	Spread of transit time (T_m) (mμsec.)
R.C.A. type 1P21 (9 stage)	5000	0·25
E.M.I. type 5311 (11 stage)	4000	4·3

7.3.3. Spread of Transit-Time in Photomultiplier Tubes.

The effect has been discussed in § 7.3.1; it is the chief factor in limiting the rise-time of the output pulse current and is due to the

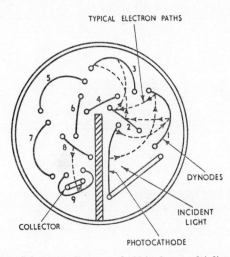

Fig. 7.4. Schematic diagram of 931A photomultiplier tube.

difference in path lengths traversed by individual electrons between electron multiplying stages. The tube should therefore be designed such that the electrons move along similar paths and the total path length, and transit-time, should be reduced as far as possible. Fig. 7.4 shows the layout of the R.C.A. type 931A photomultiplier tube which is commonly used in scintillation counters (see RAJCHMAN and SNYDER [708] and RODDA [709]). This type has nine multiplying stages and several versions have been produced using a similar dynode shape but with different photocathode structures. A more recent design employs sixteen multiplying stages (see GREENBLATT et al. [710]).

Fig. 7.5 depicts a multiplier tube employing both electrostatic and magnetostatic focussing. The electrons follow a cycloidal type path under the influence of the crossed electric and magnetic fields and the dynode multiplying surfaces are so positioned that the velocity with which the primary electrons strike the surface (about 200 electron volts) gives a high secondary emission yield. With this arrangement, SMITH [711] computes that the spread of transit-time should not exceed 0·033 mμsec.; this quantity, however, has not yet been measured. The necessity of providing a suitable magnetic field constitutes a practical difficulty in some experiments.

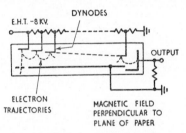

Fig. 7.5. Electron multiplier employing combined electrostatic and magnetostatic focussing.

7.3.4 Scintillation Counter Output Circuit.

The photomultiplier tube acts as a "constant" current source of high internal impedance and the output may be taken from the collecting anode or from any of the preceding dynodes. In all cases the current waveform is of the type shown in fig. 7.2. The current flows into, and is integrated by, the total stray capacitance C_1 of the tube output electrode and of the output circuit.* This capacitance is shunted by the (high) resistance R_1 in parallel with R_2 and a step type of voltage waveform is obtained, at high impedance. The output signal may then drive a cathode follower valve, so giving a similar pulse at low impedance. The rise-time of the pulse will be determined by the combined effect of the transit-time spread T_m in the multiplier and the decay time-constant T_p of the phosphor. The final amplitude will be proportional to the total number of electrons arriving at the photomultiplier anode and is thus related to the energy and type of incident nuclear particle. Care must be taken to ensure that no space charge saturation effects are taking place in

---

* The coupling capacitor C_2 is of large value and does not enter into the calculations.

the later stages of the multiplier; if such occur the gain of the tube must be lowered by reducing the E.H.T. supply voltage.

The values of the resistances R_1 and R_2 are chosen such that the charge can leak away between successive pulses, thus allowing the output voltage waveform to return to zero between events. The values should, however, not be less than that required to give an output circuit differentiating time-constant $C_1 R_1 R_2/(R_1 + R_2)$ of five times the phosphor decay time-constant. Also, in the limiting case, the time interval between successive pulses should exceed a figure equal to five times the output circuit time-constant just

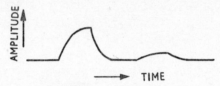

Fig. 7.6. Output pulse from delay line pulse shaping circuit.

mentioned, in order to avoid a small pulse being affected by the trailing edge of a preceding large pulse. Thus, if an anthracene crystal is used in a scintillation counter, the separation between successive pulses should not be less than 1 μsec., if reasonably accurate amplitude measurements are to be made.

When a high counting rate is to be handled, involving a time separation between pulses of less than 1 μsec., it is better to employ a delay line pulse clipping circuit than to rely on the simple resistance capacitance differentiating effect discussed above. A pulse of amplitude proportional to the photomultiplier output charge will be obtained, provided the separation between the leading edges of successive pulses is not less than ten times the phosphor decay time-constant. Fig. 7.6 shows the type of waveform obtained, and a typical pulse shaping circuit is given in fig. 7.7 (see WELLS [712]). In this circuit the resistance across the output of the multiplier tube is kept as high as possible, consistent with the mean output current not giving rise to more than about 5 V change in the mean potential of the output electrode. Each output current pulse produces a step type voltage waveform which is applied, through the cathode follower V_1, to the grid of the pulse shaping valve V_2. A voltage step of half the input amplitude travels down the cathode cable and is reflected, without change in polarity, at the open circuited end. When the return voltage step reaches the valve it raises the cathode potential and returns V_2 to its original current condition. The cathode impedance of V_2 is chosen to equal the characteristic impedance of the cable and accordingly no further reflection occurs (ideally). The output from the anode of V_2 is a negative pulse of duration equal to twice the delay-time of the cathode coaxial cable. The valve can accept another input step waveform immediately after the output pulse has terminated, without the grid potential having

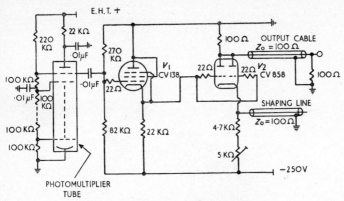

Fig. 7.7. Delay line pulse shaping circuit.
(From *Nucleonics*, Volume 10, No. 4, pages 28-30.
Copyright 1952, McGraw-Hill Publishing Company Inc.)

to return to its initial value. In practice, however, it is difficult to arrange that the cathode impedance of V_2 exactly matches the cable impedance; a second input step waveform should therefore not be applied until the lapse of about three pulse durations (six times the cable delay) after the leading edge of the first pulse, in order to allow time for the reflected step waveforms in the cathode cable to die away. A further point is that the photomultiplier output circuit, although of long time-constant, produces a small degree of differentiation of the output pulse. The ensuing "overshoot" may be balanced out by a suitable choice of the cathode resistor of V_2 which tends to produce an "undershoot" after the output pulse has passed. The action is illustrated by the waveforms shown in fig. 7.8.

We have so far considered output circuits which may be used when we require the integral of the photomultiplier output current pulse. When the measurement of time intervals is of interest it is necessary to use the leading edge of the current pulse as far as possible. The measuring circuits again require a voltage input pulse for their operation, and the rate of rise of the voltage pulse obtained from the multiplier is limited, as before, by the output stray capacitance. In this case however the shunting resistance can be quite low in value, since the amplitude of the pulse does not matter, provided the initial rate of rise of the leading edge is unaffected. For such work it is important to employ a phosphor having a very short decay-time so that the interval between the passage of the nuclear particle and the emission of the first few photoelectrons is short. Fig. 7.2 illustrates how the peak current amplitude is increased as the phosphor decay time-constant is reduced. When working

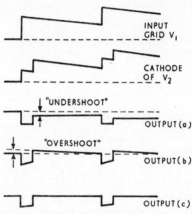

Fig. 7.8. Waveform diagram of delay line pulse shaping circuit.
(a) Input decay time-constant infinite.
(b) Cathode resistor of V_2 infinite; normal input decay time-constant (V_2 taking normal current by use of infinite negative supply voltage).
(c) Effects (a) and (b) balanced as in actual circuit.

under these conditions, the gain of the photomultiplier and following circuits should be such that the first two or three photoelectrons, which leave the photo-cathode, will provide enough current to give a useable output voltage pulse. The following typical calculation gives an idea of the value required for the gain:

Minimum output voltage pulse desired $V = 2$ V

Photomultiplier output circuit stray
    capacitance $C = 20$ pF

Output charge required $CV = 4 \times 10^{-11}$ coulombs

Therefore number of electrons required $= 2.5 \times 10^8$

Number of initial photoelectrons $= 3$

Therefore gain required $= 8 \times 10^7$

If this gain is not available in the photomultiplier, then additional amplification must be provided in the succeeding circuits. Few photomultipliers are at present obtainable with gains in excess of 10^7, but higher gain tubes should become available as a result of future development.

Care must be taken in the layout of the photomultiplier tube output connections to reduce stray reactances to a minimum. If the time of rise of the output current pulse is less than 1 mμsec. then the resistance shunting the output capacitance may be reduced to the order of 100 Ω without undue loss in amplitude; in this case the output lead may take the form of a coaxial cable sealed through the glass wall, thus eliminating the effects of lead inductance. The impedance of the coaxial line should be as high as possible.

7.3.5 Particle Counting by Čerenkov Radiation.

If a particle traverses a medium with a velocity greater than that of light in the material, then photons are emitted. These photons may be detected and amplified by a photomultiplier tube in a similar manner to the light scintillation from a phosphor. The interesting feature, however, of the ČERENKOV [713] radiation is that its decay time-constant is extremely short compared to phosphor scintillations. MARSHALL [714] states that the pulse of light has a duration of between 1 and 0·01 mμsec. If an electron travelling near the speed of light enters a block of glass with a refractive index 1·5 then he estimates that 250 photons are emitted per cm of travel. If all the photons were collected by a 5819 type photomultiplier, an average of 15·7 photoelectrons would be produced from the photocathode of the multiplier per cm travel in the glass block. A plastic material, such as Lucite, may be used instead of glass for these counters. Such counters are suited to the precise timing of high velocity particles, the accuracy only being limited by the performance of the photomultiplier tube.

7.3.6 Pulse Testing of Photomultipliers.

The photomultiplier tube is the first link in the chain of electronic circuits and it is useful to have available an artificial source of pulsed radiation for test purposes and also a circuit which can simulate the type of output pulse produced by the multiplier. One of the pulse generators described in Chapter 4 may be used for such tests, in conjunction with either of the two following injection systems.

7.3.6.1 Pulsed X-ray Tube

The type of tube is illustrated in fig. 7.9. A short duration pulse, derived from a relay or other form of pulse generator, is applied to the electron gun and a burst of X-rays is produced when the resultant electron beam strikes the target plate. A potential of 20–50 KV is normally applied to the plate and soft X-rays emerge through a very thin glass window. The pulse of X-rays of controllable duration

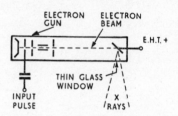

Fig. 7.9. Pulsed X-ray tube.

(1–10 mμsec. is commonly employed) occurring at a known instant of time may be used to drive the scintillation counter under test. The X-rays are weakly penetrating and a special thin window may be required to allow entry through the light shield surrounding the scintillation counter.

7.3.6.2 Pulsed Secondary Emission Valves

A valve with one stage of secondary emission may be used to simulate the dynode output pulse from a photomultiplier tube. The circuit shown in fig. 7.10 delivers the required positive pulse at high impedance. The valve is normally biassed off and is caused to pass current by applying to the grid a positive pulse, of duration 1–10 mμsec., derived from a pulse generator. The peak current required may be judged from the following figures; its magnitude may be controlled by varying the amplitude of the grid driving pulse.

Required number of output electrons = 10^8 (maximum)
Pulse duration = 1 mμsec. (minimum)
Therefore peak pulse current = 16 mA (maximum)

A small secondary emission valve is capable of delivering the required output current.

When it is desired to simulate the photomultiplier collecting anode pulse the output may be taken from the anode of the secondary emission valve; alternatively a normal pentode valve may be employed.

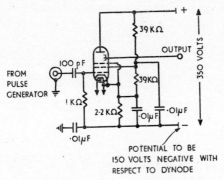

Fig. 7.10. Scintillation counter dynode output pulse simulator.
(From *Nucleonics*, Volume 10, No. 4, pages 28-30.
Copyright 1952, McGraw-Hill Publishing Company Inc.)

7.4 SPARK COUNTERS

These counters have been developed from the original work by GREINACHER [715], STUBER [716], and CHANG and ROSENBLUM [717]. KEUFFEL's [718] arrangement consists of two closely spaced parallel plates in a sealed tube with a gas filling. A high potential difference is maintained between the plates which is sufficiently large to cause a discharge when any ionization of the gas occurs. The action, which is similar to that of a Geiger-Muller tube, is as follows. When an ionizing particle enters the counter the resulting electrons form a cumulative electron avalanche which moves to the positive plate; the positive ions which are formed drift to the negative plate and a spark discharge is produced. The discharge would normally spread throughout most of the counter volume but this is prevented by applying a "quenching" pulse for a period of at least 0·05 sec. after the initial discharge. The quench waveform reduces the voltage across the counter to a sufficiently low value, such that the discharge is stopped after the first spark or electron avalanche. During the quenching period the ions are removed by the potential difference between the plates, and after this period has elapsed the counter voltage is returned to normal. The rise-time and delay-times of this type of counter have been investigated by KEUFFEL and by

MADANSKY and PIDD [719] and their results may be summarized as follows:

TABLE 7.3

Overvoltage	Output pulse rise-time ($m\mu sec.$)	Possible variation of output pulse delay ($m\mu sec.$)
250	8	15
500	–	10
900	3	3

The overvoltage is the potential difference in excess of the minimum value required for the discharge to commence.

From these figures it will be seen that such counters may be used for measurements in the millimicrosecond range. They have been applied, in particular, to the field of cosmic ray research (see ROBINSON [720]) since they are suited to experiments requiring large area counters and in which the long quench time needed is not a serious limitation.

The output pulse amplitude obtainable from this type of counter is large compared with that produced by the scintillation counters previously described.

7.5 AMPLITUDE DISCRIMINATORS

A circuit used for measuring the output pulse amplitude from a scintillation or other counter is termed an "amplitude discriminator"

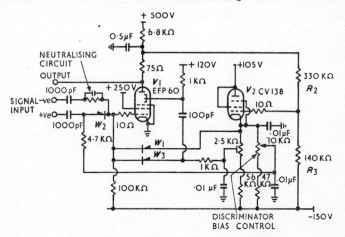

Fig. 7.11. High speed amplitude discriminator.

and forms a basic part of the counting equipment employed in nuclear physics experiments. We require a circuit which can be triggered by a short duration pulse, the amplitude of which exceeds a given voltage level; the critical amplitude should be independent of the pulse duration and rise-time, as far as possible.

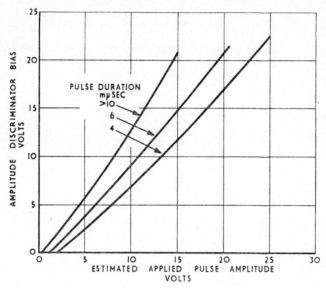

Fig. 7.12. Performance of amplitude discriminator.

The circuit consists of a non-linear element, such as a diode, which determines the critical amplitude, followed by a trigger circuit which must be sensitive to the small output derived from the preceding diode or other discriminating element. MOODY [721] has used a germanium crystal diode followed by a secondary emission trigger circuit for this purpose (see fig. 7.11).

The secondary emission valve V_1 is biassed in the centre of its normal working range. The anode current is fixed at 10 mA by the D.C. feed-back resistors R_2 and R_3; these apply automatic grid bias (negative feed-back) to the valve through the cathode follower V_2 and the crystal diode W_1. Positive feed-back is derived by capacitative coupling between dynode and grid, via the crystal diode W_3. The diode is normally biassed off by about 1 V and the attenuation introduced by it in this condition is sufficient to ensure stability in the quiescent state. Positive input pulses are applied to the grid of V_1 through the diode W_2, which is controlled by the variable bias derived from the cathode follower V_2. When the input pulse amplitude exceeds this bias, W_2 will conduct and, if the excess amplitude of the pulse is large enough, the resultant amplified positive signal at the dynode will exceed the bias on W_3. The attenuation in the dynode-grid feed-back path is then removed and the valve will drive itself hard into current. The minimum amplitude of grid pulse necessary to initiate this action is less than 0·1 V. The valve returns to its quiescent state after a period determined by the magnitude of the feed-back capacitor and various space charge saturation effects inside the valve. This period has a duration of 0·3 μsec. with the values as given in fig. 7.11. The performance of the circuit as an amplitude discriminator is shown in fig. 7.12; further relevant data are:

Discriminator amplitude range	0·05 to 25 V
Stability of triggering level	better than 0·1 V for 10 mμsec. pulses
Triggering delay	3 to 10 mμsec. dependent upon input pulse shape and drive conditions
Output pulse	200 to 300 mA
Pulse duration	0·1 to 100 μsec. readily obtainable
Output pulse rise-time	of the order of 10 mμsec.

The input discriminating diode W_2, used in the above circuit, is equivalent to a capacitance in parallel with a resistance. In the biassed-off condition the latter is high, but pulses which are smaller in amplitude than the bias voltage can still feed through the shunt capacitance and cause errors in measurement. The effect may be neutralized, if desired, by applying an additional pulse of opposite sign to the input pulse (derived from it via a phase reversing valve or transformer) to the grid of V_1, through a similar neutralizing capacitor and resistor.

In some applications it is required that the amplitude discriminator be paralysed for a comparatively long period after the circuit has once been triggered. Fig. 7.13 depicts the circuit of fig. 7.11 with the paralysis feature added (HOWELLS [722]). The valves V_3 and V_4 form a cathode coupled "flip-flop" paralysis circuit. In the quiescent condition, V_3 passes an anode current of 10 mA, which flows to earth through the diode V_5 thus keeping the cathode of V_2 at earth potential. When a positive input pulse is received, of amplitude sufficient to cause V_2 to conduct, regenerative action occurs in the latter valve, as previously described, and a large amplitude short duration pulse

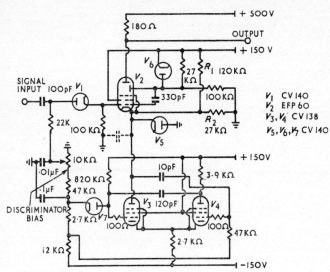

Fig. 7.13. Amplitude discriminator with paralysis circuit.

of anode current is obtained. Adequate cathode to earth capacitance must be provided in order to give the required output pulse duration. The resulting rise in the cathode potential of V_2 triggers the circuit V_3, V_4 so that V_4 now passes current and the anode current of V_3 is cut off. As soon as the cathode potential of V_2 rises above that of its grid (the rise being limited in amplitude by the diode V_6), V_2 ceases to pass current and the grid returns to earth potential. The cathode of V_2 falls to a voltage determined by the resistors R_1 and R_2 which are arranged to keep V_2 in the non-conducting condition. The discriminator circuit will then remain paralysed until the valves V_3 and V_4 return to their normal quiescent state.

Another amplitude discriminator circuit, shown in fig. 7.14, employs conventional techniques (WELLS [723]). This circuit is much slower in operation than those previously described but has a short

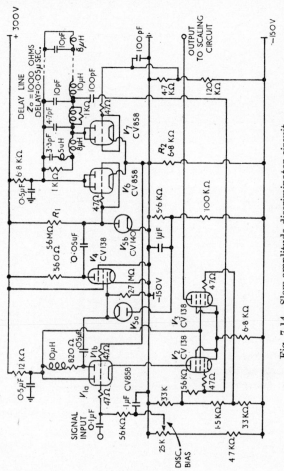

Fig. 7.14. Slow amplitude discriminator circuit.

recovery time which is determined by a delay line. In the quiescent condition, the grid of V_6 is held at earth potential by the diode V_{5b} and the resistor R_1. The grid of V_7 is at $-5\cdot5$ V so that V_6 passes all the 20 mA of current flowing through R_2; the cathode potential of V_6 will be a volt or two above earth. This cathode voltage determines the grid potential of V_{1b}; V_{1a} is held beyond cut-off by the variable discriminator bias. The cathode current of V_1 normally passes through V_2. The valves V_3 and V_4 are also biassed to cut-off.

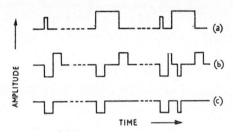

Fig. 7.15. Idealized amplitude discriminator waveform diagram.

    (a) Input pulses.
    (b) Delay line waveform (grid V_2).
    (c) V_3 anode current.

When a positive input pulse, large enough to overcome the discriminator bias, arrives at the grid of V_{1a}, current will flow in this triode, a positive signal being produced at the anode of V_{1b}. This is amplified by V_4 and a negative pulse is fed back to the grid of V_{1b} via V_6, which acts as a cathode follower. The action is cumulative and the circuit snaps over, so that all the current previously flowing in V_{1a} is transferred to V_{1b}. The grid of V_6 is driven sufficiently negative for V_7 to take all the cathode current; since the grid potential of the latter valve is fixed, the amplitude of the negative voltage step fed to the grid of V_{1b} is limited to about 6 V. It is necessary to limit this pulse in order that V_{1b} may again conduct, when the input pulse disappears from V_{1a}.

When V_7 conducts, a 10 V negative pulse, of duration $0\cdot1$ μsec., is produced across the short-circuited delay line (see fig. 7.15); this pulse is used to cut V_2 off and thus remove the current from V_1. The circuit is then completely insensitive to any further input pulse for a period of $0\cdot1$ μsec. after the triggering event. At the end of the delay line pulse, V_2 again conducts and, if the input pulse has disappeared, the circuit snaps back to its initial condition. If the duration of the input pulse is greater than 0·1 μsec. the current in V_2 will pass through V_{1a} until the input pulse voltage has fallen sufficiently to allow the circuit to return to its normal state.

As the circuit returns to normal, V_7 will be cut off and the delay line will now produce a positive 0·1 μsec. pulse; when this pulse has lapsed the arrangement is back in its quiescent condition. The circuit may, however, be retriggered before the quiescent condition is completely reached, but the duration of the output pulse is then reduced and may fail to operate the succeeding scaling circuit. When used in conjunction with the scaling unit of fig. 7.17 a dead-time of 0·25 μsec. was obtained, as determined by a random pulse experiment. Deterioration in waveforms accounts for the increase in the observed dead-time over the expected value of between 0·15 and 2·0 μsec.

The minimum pulse duration required for the circuit to operate is about 50 mμsec.; this figure should be contrasted with the value 6 mμsec. which applies to the previous arrangement shown in fig. 7.12. The circuit of fig. 7.14 could probably be improved in this respect by the use of secondary emission valves.

7.5.1 Use of Amplitude Discriminators.

The applications of amplitude discriminators in nuclear physics experiments fall into two general classes:

> (a) The scaling of nuclear counter pulses above a given amplitude, so that the number of nuclear particles of energy greater than a certain amount may be determined. In such cases the discriminator must drive a scaling circuit and, if a high counting rate is to be used, a fast amplitude discriminator must be employed. The duration of the pulse derived from the nuclear counter is usually under our control and should be determined by a consideration of the counting rates employed. It should usually be made as long as possible, consistent with one pulse having decayed to zero before the next pulse arrives; in general, the pulse duration would be kept to the same order of magnitude as the dead-time of the scaling and amplitude discriminating circuits. In particular, if a circuit of the form shown in fig. 7.7 is used, the duration of the input pulse to the discriminator need not be less than one quarter of the amplitude discriminator dead-time.

(b) The examination and measurement of the amplitudes of and time intervals between pulses which are above a given amplitude. The principal requirement which an amplitude discriminator circuit must fulfill for these applications is that there shall be a minimum time delay in triggering; the circuits of figs. 7.11 and 7.13 are therefore preferable to that of fig. 7.14. The dead-time of the circuit may not be of much importance in such applications. The circuit of fig. 7.13 was designed for use in the recording oscilloscope of fig. 6.15, where it was necessary to include a long controllable paralysis-time in the discriminator.

7.6 FAST SCALING CIRCUITS

A further circuit which is required, when nuclear events are to be counted, is the scaling unit. This circuit scales down the number of pulses derived from the output of the preceding amplitude discriminator circuit. The reader who is not familiar with nuclear physics should realize that the pulses to be counted occur at random instants in time, so that two pulses may occasionally be produced with a very small time separation between them, whilst at other times, this separation may be large.* Scaling circuits differ from frequency dividing circuits in that they must be completely aperiodic, and must start and stop scaling when a control switch is operated. In the case of random pulses, the possible time separation may approach zero, so some pulses will inevitably be lost owing to the dead-time of the circuits. If the dead-time is short enough, however, only a small percentage of the pulses will fail to be counted and the counting loss can be estimated if the circuit dead-time is known. A 1% counting loss can be tolerated in general and, in some cases, counting rates giving a loss as high as 10% may be acceptable.

Many circuits have been devised for scaling pulses whose time separation is not less than 2 μsec. An average counting rate of 5000 per sec. may then be scaled for a 1% counting loss. If the circuit dead-time is reduced to 0·2 μsec. then an average rate of 50,000 per sec. may be scaled or 500,000 per sec. if a counting loss of 10% can be tolerated. It should be noted that a scaling unit is often preceded by an amplitude discriminator and, in the case of

---

* In order to obtain a reasonably accurate measure ($\sim$ 1% error) of the average number of pulses arriving per second, a minimum of 10,000 pulses total should be counted.

fast circuits, the dead-time of the discriminator is usually greater than that of the scaling circuit.

Two types of fast scaling circuit are in use to date; the first to be described is a development from the Eccles-Jordan low speed scale-of-two unit. The arrangement, designed by FITCH [724], is shown in fig. 7.16. The triodes V_3 and V_4 form the normal direct coupled Eccles-Jordan binary circuit, except that the cathode follower valves V_2 and V_5 have been inserted in the anode to grid feed-back paths. The cathode followers reduce the capacitative loading on the anode

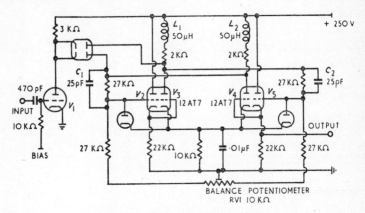

Fig. 7.16. Fast scale-of-two circuit.

to grid coupling circuits to a minimum and thus allow capacitors C_1 and C_2 also to be reduced to a minimum (consistent with the duration of the input trigger pulse). The anode to grid capacitance of the valves V_3 and V_4 may be neglected, since the grids are driven from the low output impedance of the cathode followers.

The circuit has two stable states in which either V_3 or V_4 passes anode current; it may be triggered from one state to the other by the application, through the coupling diodes, of a short duration negative pulse derived from the anode of V_1. The trigger pulse must not be larger than one third of the total change in voltage across the anode load of either V_3 or V_4 and its duration should be short compared with the "memory" time-constant of the circuit. The latter quantity is the time interval required for the circuit to change from the condition of V_3 conducting to V_3 cut off, or vice-versa, and for the circuit reactances to approximately attain their quiescent state conditions after the change over in valve currents. These reactances consist of C_1, C_2, L_1, L_2 and the stray capacitances to earth. For correct operation it is essential that the circuit be balanced; the potentiometer RV_1 is provided for this purpose so that, in their respective conducting states, the anode currents of V_3 and V_4 may be adjusted to be equal.

If the circuit is balanced in this manner, a continuous input pulse recurrence rate of 15 Mc/s may be counted, whilst on random input

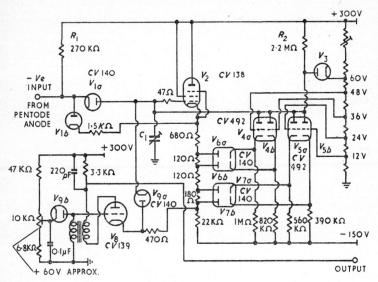

Fig. 7.17. Scale-of-five circuit.

pulse operation the circuit dead-time, after counting each pulse, is less than 0·1 μsec. It should be remembered that in all cases the input trigger pulse duration should not be greater than 0·05 μsec. (preferably rather less) whilst some precautions must be taken to limit the amplitude of the trigger pulses.

A similar circuit using pentodes instead of triodes, to reduce anode to grid capacitance, has been described by SESSLER and MASKET [725].

One such binary circuit may be followed by $(N-1)$ other similar binary units thus giving a scale of 2^N. It is difficult to design a scale-of-ten circuit on these lines whilst still maintaining a high counting speed so that it is common practice to employ a scale-of-eight followed by normal types of low speed decade scalers.

The second type of circuit which is in use provides a scale of ten. It consists of a scale-of-five followed by a binary stage, the basic circuit of the scale-of-five, designed by WELLS [723], being as shown in fig. 7.17.

The arrangement is a development of a familiar frequency dividing circuit, but complete aperiodicity is achieved by the addition of valves V_4 to V_7. This scaler was originally designed for use in conjunction with the discriminator circuit of fig. 7.14 and its speed is sufficiently high to count all the output pulses from the latter (which has a dead-time of 0·25 μsec.).

Negative input pulses, of 0·1 μsec. duration, are derived direct from the anode of a pentode valve (with anode load resistance to H.T.) which supplies a pulse current of 8 mA. This current pulse is fed into the capacitor C_1 through the diode V_{1a} and the value of C_1 is adjusted such that a voltage step of 12 V amplitude is produced. The total capacitance required, including all stray capacitances, is about 70 pF. During the drive period the diode V_{1b} is cut-off by the grid to cathode voltage of the cathode follower V_2. When the input current pulse ceases, the resistance R_1 causes the cathode potential of V_{1a} to rise until arrested by the diode V_{1b} at a volt or two above the cathode potential of V_2. The diode V_{1a} is thus cut-off, leaving the capacitor C_1 isolated. By this process the capacitor is discharged in successive 12 V steps, one for each input pulse.

C_1 is initially charged to a potential of 60 V so that on the fifth input pulse the voltage falls to zero and the blocking oscillator valve V_8 is brought into conduction by the cathode follower V_2. V_8 then drives itself into current and the cathode and grid potentials rise rapidly, thus recharging C_1 to its initial level through the diode V_{9a}. The rise in potential of the grid of V_8 is arrested by the diode V_{9b} at a voltage which is adjusted such that C_1 is recharged to exactly 60 V. It should be noted that, as soon as the cathode potential of V_8 rises above that of the grid, the anode current starts to diminish; the action is again cumulative and the grid swings rapidly back to earth potential. The cathode voltage also falls, thus cutting off the diode V_{9a} and leaving the capacitor C_1 free to accept further input pulses. The voltage waveform observed across C_1 is sketched in fig. 7.18.

It has been previously mentioned that valves V_4 and V_7 are included to make the circuit completely aperiodic; the action of this "capacitor locking circuit" is as follows. Initially, the potential difference across C_1 is held at 60 V by the resistor R_2 and the diode V_3 which is connected to a 60 V tapping point on the potentiometer chain. The four diodes V_6 and V_7 are all conducting and the four triodes V_4 and V_5 are cut off. On the arrival of the first input pulse the potential across C_1 is driven down to 48 V and during this step the diode V_{6a} cuts off, leaving V_{4a} passing current. By virtue of the high anode load R_2 the latter triode holds the capacitor to within a volt or two of its grid potential which is fixed at 48 V. When the second pulse occurs the potential of C_1 is depressed to 36 V, and during

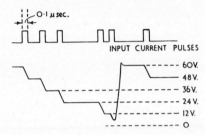

Fig. 7.18. Scale-of-five circuit. V_2 grid waveform.

this step the diode V_{6b} cuts off leaving V_{4b} passing current. V_{4b} then maintains the capacitor at its grid potential of 36 V, the anode current of V_{4a} being suppressed because its anode potential is now below that of its cathode. Triodes V_{5a} and V_{5b} function in a similar manner on the third and fourth pulses and lock the capacitor voltage to the values 24 and 12 V respectively. This circuit action is performed at a comparatively low speed, so that, if a high input pulse rate is being scaled, the locking circuit will not have time to operate; the scaler continues to function nevertheless and the influence of the locking circuit may be ignored. At low input rates, on the other hand, the locking circuit holds the capacitor voltage fixed between input pulses, so that the effects of all leakage currents associated with the capacitor are removed.

One output pulse is obtained from the blocking oscillator V_8 for every five input pulses and this may be used to drive a conventional scale-of-two circuit, giving a scale of ten for the complete instrument.

The blocking oscillator V_8 may be replaced by the trigger circuit of fig. 4.9; pulses with a separation of less than 0·25 μsec. (the nominal design figure for the above arrangement) may then be counted. If pulses occurring at a regular high recurrence rate are to be scaled, then the capacitor locking circuit V_4 to V_7 may be omitted.

7.7 COINCIDENCE CIRCUITS

The function of a coincidence circuit is briefly as follows. Two nuclear counters are used giving two sets of input pulses which drive the coincidence unit; the circuit is then required to give an output pulse every time a pulse from one counter arrives within a given short time interval of a pulse from the other counter. This time

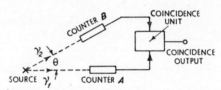

Fig. 7.19. Angular correlation experiment.

interval is termed the "coincidence resolving time". Circuits of this nature, having coincidence resolving times of 1 μsec. and longer, have been in use for many years, but with the development of the scintillation counter, described in § 7.3, it has been found possible to design circuits having resolving times as short as 2 mμsec.

A typical experiment, which illustrates the use of a coincidence circuit and the importance of having a short stable resolving time, is the measurement of the angular correlation of the γ-radiation from a radioactive nucleus. Certain radioactive nuclei emit two γ-rays almost simultaneously and it is of interest to determine the rate of disintegration for different angles between the γ-ray emission directions. Two directional counters are used, as shown in fig. 7.19, and the coincidence circuit gives an output pulse every time a nucleus emits the two γ-rays in the appropriate directions (counter inefficiencies are ignored).

Let N be the average number of nuclear disintegrations occurring per second and let the two types of γ-ray be denoted by γ_1 and γ_2. The number of output pulses from counter A due to γ_1 quanta is then Ne_1 where e_1 is the efficiency of the counter involving, among other factors, the solid angle subtended by the counter at the source. A few of these particular nuclei will give rise to a γ_2 quantum travelling in the direction of counter B so that a coincidence output counting rate of $Ne_1e_2W(\theta)$ will be obtained, where e_2 is the efficiency of counter B and $W(\theta)$ is a function of the angle between the counters. If there is no preferred angle between the two γ-rays then $W(\theta)$ is unity. Now counter B may also register a γ_1-ray and give a coincidence output with a γ_2-ray passing through counter A hence we have:

Total coincidence output counting rate $= N_c$ say

$$= 2Ne_1e_2W(\theta) \qquad (7.7)$$

(assuming that the counter efficiencies for the two types of γ-ray are the same).

In addition to these genuine coincidences, some output signals will be produced due to chance coincidences between pulses from the two counters. We therefore have

$$N_r = 2N_A N_B \tau \qquad (7.8)$$

where N_r = random coincidence output rate,

N_A = average number of pulses from counter A per unit time,

N_B = average number of pulses from counter B per unit time,

τ = coincidence resolving time.

The factor 2 in this relation arises because the coincidence unit operates if a pulse from B arrives within the time interval τ of a pulse from counter A, irrespective of which pulse occurs first.

Now

$$N_A = 2Ne_1 \; (\gamma_1 \text{ and } \gamma_2 \text{ quanta})$$
$$N_B = 2Ne_2 \; (\gamma_1 \text{ and } \gamma_2 \text{ quanta})$$

therefore by relation 7.8

$$N_r = 8N^2 e_1 e_2 \tau$$

Using equation 7.7 we thus have

$$\frac{N_c}{N_r} = \frac{1}{4N\tau} W(\theta) \qquad (7.9)$$

Now it is necessary to keep N as high as possible, so that the true coincidence rate is large enough to give a sufficient number of counts (10,000) for adequate accuracy in a reasonable experimental time. This rate is proportional to the solid angle of collection by the counters, so that, to achieve good angular resolution, N must be increased to a maximum in order that this solid angle may be kept small. The experiment will not be successful, however, unless the ratio N_c/N_r is greater than 10, say. Relation 7.9 shows that τ, the coincidence resolving time, must accordingly be reduced to a minimum. It should be noted that N may be controlled by varying the amount of radioactive material used in the experiment.

The reader should also note that, since N_r is a function of τ, the coincidence resolving time must be constant during the course of the experiment in order that a correction may be made to the experimental results to allow for the presence of random coincidences. In an actual experiment the total coincidence output rate will be equal to $(N_c + N_r)$; N_r is measured as described in § 7.7.1 and the value subtracted from the total, thus giving the desired quantity N_c. The experiment is repeated for different angles θ between the two counters hence enabling the function $W(\theta)$ to be determined.

In some experiments the number of counter output pulses in each separate channel will greatly exceed the values $2Ne_1$, $2Ne_2$ above, due to the presence of radiations or particles other than the two relevant related γ quanta. The need for the shortest possible resolving time is further emphasized, since N_r is increased without any corresponding increase in N_c.

Various forms of coincidence circuit will now be considered in detail.

7.7.1. Pulse Limiter with Diode Mixer Circuit.

An arrangement which has been extensively used by BELL et al. [726], WELLS [727], and other workers, is shown in fig. 7.20; negative input pulses having fast rising edges, followed by a slow exponential decay to zero, are required.

Each of the valves V_1 and V_2 normally passes an anode current of 20 mA. If an input pulse has sufficient amplitude one valve will be completely cut off, thus driving a current of 20 mA into the anode circuit, which consists of a matched 100 Ω coaxial cable. This anode pulse travels down the cable to the T-junction where the current divides. Part* flows down the 50 Ω cable to the short-circuited end where the voltage is reversed in phase, and returns to the T-junction. When it arrives back at the junction no further reflection occurs since the two 100 Ω lines match the 50 Ω cable. The pulse then splits into two "halves" which continue down the 100 Ω cables to their ends

---

* The remaining portion travels along the other 100 Ω line and is absorbed in the terminating resistor.

and are absorbed in the terminating resistors. The nett result is that a large negative input pulse applied to the grid of either V_1 or V_2 will give a positive pulse of 0·5 V amplitude at the cable T-junction

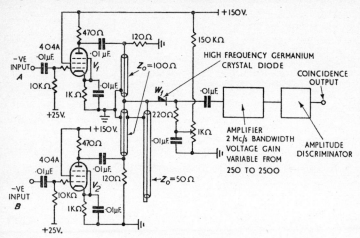

Fig. 7.20. Fast coincidence circuit.

the duration of which is determined by twice the delay-time of the 50 Ω cable. This pulse is delayed after the input signal by the transit time of one of the 100 Ω connecting cables. If two input pulses arrive

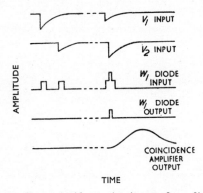

Fig. 7.21. Fast coincidence circuit waveform diagram.

simultaneously the resultant signals will add at the T-junction and give a pulse of amplitude 1 V. The 1 V pulses, due to coincidences, are sorted out from the non-coincident 0·5 V pulses by the crystal diode which is suitably biassed. Waveforms are shown in fig. 7.21.

The output from the coincidence diode is of small amplitude (the duration being determined by the overlap time of the two anode pulses) and feeds into an amplifier which can be of comparatively narrow bandwidth. The 0·5 V amplitude non-coincident pulses give rise to a small output due to the backward resistance and shunt capacitance of the crystal diode. If the amplitude of the amplifier

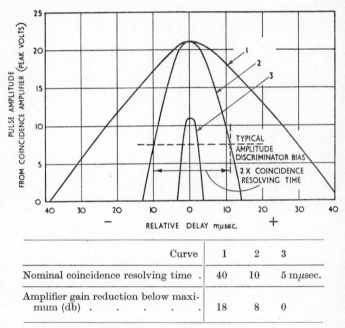

	Curve	1	2	3
Nominal coincidence resolving time .		40	10	5 mµsec.
Amplifier gain reduction below maximum (db)		18	8	0

Fig. 7.22. Fast coincidence circuit performance under test pulse conditions.

output signal is measured as the time separation between the two input pulses is varied a curve of the form shown in fig. 7.22 is obtained. The coincidence diode bias is adjusted to give the maximum ratio between the coincident and non-coincident outputs; the ratio should not be less than 5. The gain of the low frequency amplifier is set such that the maximum output for coincidence lies between 10 and 20 V, and the succeeding amplitude discriminator is arranged to trigger at about one third of this level. It should be noted that the setting of this discriminator will effectively vary the coincidence resolving time, so that the bias level should be stable and the gain of the amplifier should also be well stabilized.

The coincidence resolving time may again be varied (as shown in fig. 7.21) by altering the duration of the shaped pulses from V_1 and V_2; this duration is controlled by the length of 50 Ω cable employed. The gain required in the low frequency amplifier will need to be increased as the resolving time is reduced, owing to the shorter pulse output delivered by the coincidence diode.

It will be seen that the coincidence resolving time is equal to the duration of the pulses at the input to the coincidence diode, less the

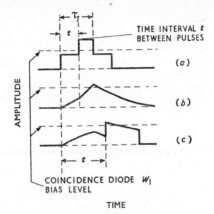

Fig. 7.23. Typical waveforms at input to coincidence diode in fast coincidence circuit.

time overlap required to give an integrated signal through the diode sufficient to operate the following amplifier and discriminator circuits. So far, we have assumed that the pulse shapes are perfectly rectangular, but, in practice, waveforms such as those depicted in fig. 7.23 will be observed owing to the finite rise-time of the input pulses applied to the grids of the limiting valves V_1 and V_2. Three typical waveforms are illustrated:

(a) Both input pulses are rectangular and of duration τ_1; we shall assume that the coincidence unit gives an output if the sum of the two pulses exceeds the bias level shown in the figure. The maximum time separation between the leading edges of the pulses to give a coincidence output will then be $\pm \tau_1$, i.e. the coincidence resolving time is simply τ_1.

(b) Both input pulses are similar in shape, but with a slow rate of rise and fall. In this case the maximum allowable time separation between pulses, for a coincidence output to be given, will be considerably less than τ_1, so that the coincidence resolving time is smaller than τ_1. If the rise-time of the pulses is rather longer than τ_1, then the sum of the two pulses, even when directly superimposed, may be insufficient to operate the coincidence circuit, i.e. the resolving time is zero.

(c) One input pulse is rectangular whilst the other has a slow rate of rise and fall. In this case, if the fast pulse follows the slow one, then the maximum time between the leading edges of the two pulses may exceed τ_1 before the coincidence output is lost, as shown in fig. 7.23(c). If, on the other hand, the fast pulse precedes the slow one then this maximum time separation will be less than τ_1. The coincidence resolving time may accordingly be either greater or less than τ_1.

The values predicted for the resolving time will all require to be modified in order to allow for the time overlap previously discussed.

The input pulses in an actual nuclear physics experiment will normally be derived from a scintillation counter and a large range of input pulse amplitudes will be met with. The time taken for these pulses to cut off the input valves V_1 and V_2 will vary with their rates of rise so that the rise-times of the pulses applied to the coincidence diode will according vary from pulse to pulse. In particular, input pulses of small amplitude will give slow rising pulses as in (b) above.

Now let us consider two scintillation counters arranged to detect coincident nuclear radiation and let a variable delay line be placed in series with the output pulses from one counter. When the delay is zero, the coincidence circuit output will be derived from all input pulses which give a sufficiently short rise-time at the coincidence diode and which therefore have a finite coincidence resolving time. As the delay time is increased, only those input pulses which can produce a resolving time in excess of the delay will contribute to the coincidence output. The mean coincidence output pulse rate will therefore fall because true coincidences between very low amplitude input pulses will be lost as in (b) above. When the delay inserted becomes greater than the nominal pulse width, some coincidence output pulses will still be observed due to coincidences between pairs of very large and small amplitude pulses (see (c) above). Typical experimental curves of the variation of the average coincidence output pulse rate with delay are shown in fig. 7.24 where positive and negative delay indicates that the latter is inserted in series with counter A or counter B respectively. If the delay is sufficiently large, then all true coincidences are lost and the residual counts are due to purely random coincidences. It should be noted that the introduction of a long delay line in this manner gives an experimental method of determining the random coincidence rate in a particular experiment.

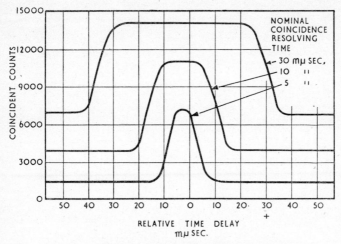

Fig. 7.24. Experimental characteristic curve for fast coincidence circuit. (Including use of amplitude discriminators as in § 7.7.2.)

The rate should be independent of the delay, provided that the introduction of the delay does not indirectly change the coincidence resolving time.

The curves of fig. 7.24 indicate the performance of a coincidence unit in a particular experiment; a consideration of these graphs will show the minimum resolving time that can be employed, without an undue loss of coincidence counting efficiency at the maximum of the curve.* The delay line used was in the form of a variable length of coaxial cable which could be switched into circuit; this cable is an essential part of a practical coincidence unit and enables the curve of coincidence rate versus delay to be measured conveniently at the start of an experiment.

---

* The maximum may not occur at zero delay if there are unequal transit-times in the two scintillation counters or if there is a small time difference between the "coincident" nuclear radiations or particles being detected.

From the above discussion of the influence of pulse shapes on the coincidence resolving time, it becomes apparent that the following two conditions must be met, if the minimum possible coincidence resolving time is to be obtained without undue loss of counting efficiency:

(a) The rise-time of the pulses at the coincidence diode must be as short as possible, so that their duration may be reduced to a minimum. As an alternative to reducing the pulse duration, the circuit following the diode may be arranged to require a considerable amount of pulse overlap, before a final coincidence output signal can be produced.

(b) Only those counter pulses which are due to the incident particles under investigation should, as far as possible, be allowed to operate the coincidence circuit. All other pulses may produce unwanted coincidence outputs and often give rise to some deterioration in the performance of the unit.

Condition (a) implies that the input signals from the counters should cut off the limiter valves as rapidly as possible; their amplitudes must therefore be large. Adequate current gain must be present in the circuit between the photocathodes of the multiplier tubes and the coincidence limiter valves, such that the first one or two photoelectrons released by the light scintillation will provide sufficient signal (say 2 V at the grids of V_1 and V_2). Typical values are quoted in § 7.3.4.

In order to meet condition (b) full use should be made of the fact that the wanted pulses from the counters may differ in amplitude from the noise and unwanted pulses. Discriminator circuits may be introduced into the coincidence unit so that such an amplitude selection may be made; BELL's [726] technique, as described in the next section, may be employed.

The reader should note that when a high value of photomultiplier gain is employed in order to obtain optimum coincidence resolution then a considerable number of noise pulses will be present in the counter output. These pulses are large enough to give rise to a random coincidence output but are generally considerably smaller than those produced by the light scintillations from the phosphor. It is therefore essential to include amplitude discrimination, so that all coincidence output pulses derived from small amplitude noise signals are suppressed.

7.7.2 Pulse Amplitude Selection for Coincidence Units.

Preliminary selection is achieved, as far as possible, by using suitable filters or by employing spectrometric methods in front of the scintillation counter. A mixture of incident radiations may still enter the counters and some further discrimination may be obtained by electronic pulse amplitude selection after the counter. Such discrimination will satisfy condition (b) above and give a good measure of protection against photomultiplier noise.

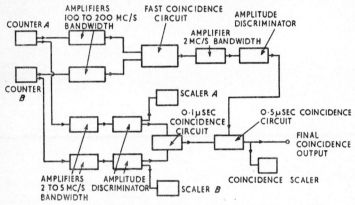

Fig. 7.25. Block schematic diagram of complete fast coincidence unit.

A block schematic diagram of a complete coincidence unit, embodying amplitude discrimination facilities, is given in fig. 7.25. The pulses from the photomultiplier tubes are fed to amplitude discriminators, after some amplification if necessary. The amplifiers can be of comparatively narrow bandwidth (a figure of 2 to 5 Mc/s being commonly used) determined by the input pulse duration.

The amplitude of the pulse taken from the scintillation counter for this purpose must be proportional to the total number of electrons leaving the photocathode of the multiplier for each scintillation detected. It is sometimes necessary to utilize the pulse from a dynode, previous to the final collecting anode of the multiplier, in order to avoid space charge saturation effects in the last stages of electron multiplication (see BEGHIAN et al. [728]).

The outputs from the two amplitude discriminators are taken to a subsidiary coincidence circuit possessing a resolving time of about 0·1 to 0·5 μsec. and the output of this circuit is used to gate the fast coincidence circuit output. A final output pulse can therefore only be obtained from two input pulses the amplitudes of which are greater than the discriminator settings and the time difference between which is less than the resolving time of the fast coincidence circuit. The amplifier gains are adjusted such that a pulse which has

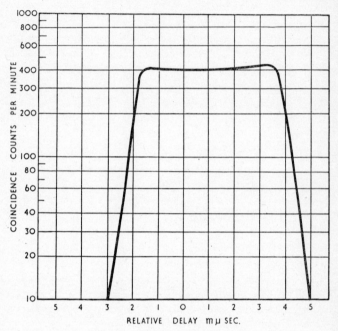

Fig. 7.26. Response of fast coincidence circuit with pulse amplitude selection.

an amplitude equal to the discriminator level will be many times larger than the minimum value required to operate the high speed coincidence circuit (see § 7.7.1). Scaling units may also be introduced to count the pulses from the amplitude discriminators; the average pulse rates, in each channel, which contribute to the final coincidence rate may then be determined.

This technique may be extended by employing pulse height selectors instead of simple amplitude discriminators, such selectors having an upper as well as a lower limit of amplitude discrimination. This type of circuit is only operated by pulses the amplitudes of which lie within the prescribed range.

When experimental conditions allow of such a drastic reduction in counting rate, the latter method may be employed to advantage to improve the fast coincidence circuit performance. Since the amplitudes of the pulses being fed to this circuit, which give rise to a final output, are restricted to a narrow range, the shape of the pulses at the coincidence diode will be similar and, in this special case, the coincidence resolving time may be reduced below the rise-time of the latter pulses. It should be noted that it is always possible to detect small time differences between pulses of similar shape and amplitude even if the time differences are much less than the pulse rise-time.

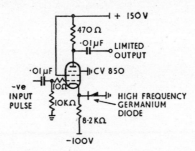

Fig. 7.27. Current stabilized limiter circuit.

Fig. 7.26 shows the coincidence delay curve obtained by McLusky and Moody [729], using this technique.

The above indirect method of introducing amplitude discrimination to coincidence circuits is necessary, because the fast coincidence unit must be operated by the pulse derived from the first few photoelectrons produced in the photomultiplier tube, whereas the amplitude discriminator requires to be driven by a pulse proportional to the total number of photoelectrons released by the light scintillation from the phosphor. The discriminator output pulse will therefore be considerably delayed with respect to the pulse operating the fast coincidence unit and, moreover, this delay will vary with pulse amplitude. The fast coincidence circuit therefore cannot follow the amplitude discriminator, as is the case in more conventional low speed circuits.

7.7.3 Stability of Coincidence Circuits.

As pointed out earlier the coincidence resolving time must be stable in value. The anode currents of the input limiter valves V_1 and V_2 (fig. 7.20) should accordingly be stabilized and the arrangement of fig. 7.27 will be found useful in cases where the mean input pulse rate is high. In this circuit the cathode current of the valve is diverted to earth, through the crystal diode, when the input pulse is applied. It should be realized that if this diode is replaced by a large capacitor, as in fig. 7.20, the mean cathode current of the valve is stabilized but the peak anode current will be affected by the input pulse rate. This latter circuit is satisfactory for low input rates, but that of fig. 7.27 should be used if the duty cycle (mark to space ratio) of the input pulses is greater than say 5% (1 : 20).

Another source of instability is to be found in the germanium crystal coincidence diode. The backward resistance of such diodes varies considerably with temperature and so the effective bias may be altered when a relatively high value resistive diode load is used. The load resistance should not exceed a few hundred ohms and the temperature of the crystal must be kept constant.

One other difficulty encountered is that of keeping the coincidence resolving time unaffected by change in delay cable length. The delay variations usually lead to small changes in resolving for the following reasons:

(a) For long delays, the attenuation in the cable causes a deterioration in the shape of the pulse applied to the coincidence diode.

(b) For short delays, the small mismatches at the cable input ends give rise to reflected pulses which arrive back at the coincidence diode before the initial pulse has passed, thus altering the effective shape of the latter.

Effect (a) may be minimized by inserting the delay before the limiter valves, if possible. Effect (b) may be avoided by keeping the delay greater than about one half the coincidence resolving time; only the first reflected pulse is troublesome since subsequent reflections are of negligible amplitude.

To summarize, coincidence resolving time stabilities of $\pm 5\%$ are readily obtainable, except on the very shortest resolving time settings. If this figure is to be improved upon, great care must be exercised is stabilizing all components in the system.

7.7.4 Factors Determining the Minimum Possible Resolving Time.

There are two limitations which prevent the resolving time of a coincidence circuit from being decreased below a certain

APPLICATIONS TO NUCLEAR PHYSICS

minimum value, if the coincidence counting efficiency is to be kept high.

(a) As discussed previously (§ 7.7.1), the time taken for the counter output voltage pulses to cut off the grids of the limiter valves (fig. 7.20) determines the minimum resolving time of the circuit. To reduce the rise-time to a minimum a large gain must be available in the photomultiplier and subsequent amplifier, and the rise-time is then determined by the spread of transit-time of the photomultiplier tube rather than by the decay time-constant of the phosphor. We have seen how the coincidence resolving time may be reduced to a value somewhat below that of the rise-time of the pulses by employing pulse amplitude selection, but it is found that the effect of the spread in transit-time is not the same for every pulse, so that the rise-times of the output pulses will still vary thus determining a minimum useable coincidence resolving time. This effect is less than 1 mμsec. with the R.C.A. type 937A photomultiplier and so may generally be neglected; a value of several milli-microseconds obtains, however, with the E.M.I. type 5311 tube.

(b) A time delay exists between the passage of a nuclear particle into the phosphor and the emission of the first photo-electron from the photomultiplier cathode. This delay is subject to random variations and therefore sets a lower limit to the useable coincidence resolving time.

Post and Schiff [702] have shown that the mean value τ_0 of the time delay is given by

$$\tau_0 = \frac{T}{R}\left(1 + \frac{1}{R}\right), \quad R \gg 1 \qquad (7.10)$$

where T is the mean life-time of the light flash from the phosphor and R is the total number of photoelectrons produced by the flash.

The coincidence resolving time of the circuit should be set at a value greater than $3\tau_0$ if the probability of observing the first photo-electron is not to be less than 0·95.

Fig. 7.28 shows Bell's curves illustrating this effect; the minimum resolving time is plotted in terms of the electron energy expended in the crystal. For a stilbene crystal and electrons with an energy of 100 KeV the minimum resolving time is 1 mμsec.

The minimum coincidence resolving time that has been obtained in practice is in good agreement with the value set by the two limitations just discussed. The time limitation (b) above is of much less importance in experiments on high energy particles using

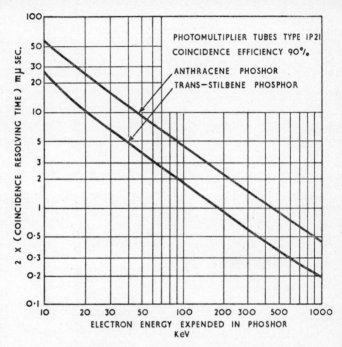

Fig. 7.28. Theoretical minimum coincidence resolving time.

Čerenkov counters (§ 7.3.5) than when a phosphor is employed to detect low energy particles. BAY et al. [730] report that resolving times of less than 1 mμsec. may be attainable in the former case.

7.7.5 Mixer Circuits.

The performance of the coincidence circuit described in § 7.7.1 has been considered in detail because it is of a commonly used type, and a study of the pulse shapes and time delay variations is applicable in general to most other forms of coincidence circuit. Some alternative arrangements will now be briefly reviewed. Pulse amplitude selection may be applied to all by using the technique described in § 7.7.2.

7.7.5.1 Crystal Diode Circuits

BAY [730], [731], BALDINGER et al., [732] and ELLMORE [733] have used crystal diode mixing circuits, driven directly by the current

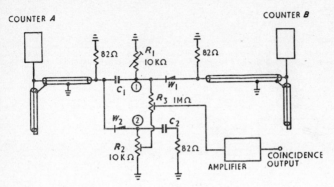

Fig. 7.29. Crystal diode coincidence circuit.

output pulses obtained from scintillation counters. We then have examples of low capacitance circuits being operated at rather lower amplitude levels than other types of mixing circuit employing

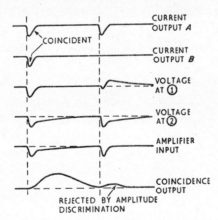

Fig. 7.30. Waveform diagram of diode coincidence circuit.

thermionic valves. One circuit designed by BAY is shown in Fig. 7.29 and its operation may be understood by a consideration of the waveform diagrams in fig. 7.30.

Current pulses from the photomultiplier tube are fed into the cables thus producing negative voltage pulses at the cable outputs. The duration of these pulses may be controlled by varying the lengths of the short-circuited pulse shaping lines, the maximum duration being determined by the shape of the current pulse as illustrated in fig. 7.30. Single pulses from counter A will charge the two capacitors C_1 and C_2 through the diodes W_1 and W_2. The rates of discharge of the capacitors are equalized by adjustment of R_1; the difference between the voltages across C_1 and C_2 is applied to a low frequency amplifier. The potentiometers R_2 and R_3 may be adjusted such that there is negligible output from the amplifier for single A pulses; the balance is maintained over a large range of pulse amplitudes. Single negative pulses from counter B will have little effect since diode W_1 will be cut off. However, when a pulse from B occurs at the same time as one from A, then the resultant charge on the capacitor C_1 will have a value different from that which obtains when the B pulse is absent. The balance for the A pulse is therefore disturbed and the amplifier produces an output. The coincidence resolving time is controlled by the pulse durations and so may be varied by altering the lengths of the pulse shaping cables. For the diode to operate efficiently, the input pulse amplitudes should not be too small, and BAY has given the following minimum gain requirements which the photomultiplier tubes must fulfil:

TABLE 7.4

Resolving time ($m\mu sec.$)	Minimum gain (no. of of electrons at tube output)
1·85	$3·5 \times 10^7$
0·54	6×10^7
0·29	16×10^7

The above figures for the gain are based on the supposition that the circuit is to respond to the current derived from a single photoelectron leaving the photocathode of the scintillation counter.

BALDINGER, et al. [732], have also described a crystal diode coincidence circuit consisting of a diode bridge which is balanced for non-coincident pulses but unbalanced for coincident ones. ELLMORE [733] has discussed a lower speed crystal diode circuit.

Another crystal diode arrangement, which is driven directly by pulses from the two counters, but which depends on a somewhat different principle for balancing out the effect of non-coincident pulses will now be described.

7.7.5.2 Double Pulse Technique

The schematic diagram of a circuit used by DICKE [734] is given in fig. 7.31. The pulse from counter A travels along the coaxial cable l_1 to the cable junction where the current divides between

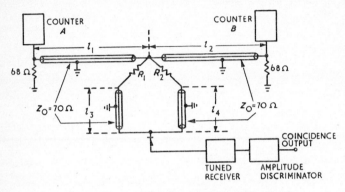

Fig. 7.31. Double pulse coincidence circuit.

the three other lines. The resistors R_1 and R_2 are carefully matched to equalize the currents which flow down cables l_3 and l_4, so that pulses of equal amplitude arrive at the diode with a time separation equal to the difference in delays of the two cables i.e. proportional to the lengths $l_4 - l_3$. The pulse from l_3, after passing the diode, will continue down l_4 until it reaches the matching resistor R_2 where it is absorbed. The network comprised by R_1, R_2, l_1 and l_2 is designed such that both the lines l_3 and l_4 are correctly terminated, whilst the attenuation suffered by a pulse travelling from l_4 through R_2 and R_1 into cable l_3 is at least 20 db. The two equal pulses from the diode are fed into a narrow-band radio receiver which is tuned to give a minimum output pulse (after detection). The lowest frequency at which this occurs is when the difference in time delay of the cables l_4 and l_3 is equal to half the period of the receiver frequency i.e.

$$f_0 = \frac{1}{2(T_4 - T_3)} \qquad (7.11)$$

where f_0 = frequency to which receiver is tuned
T_3 = time delay of cable l_3
T_4 = time delay of cable l_4

When the receiver is set to this frequency, and provided the two pulses from l_3 and l_4 are exactly equal in amplitude and shape, a very small output will be obtained for single pulses from counter A.

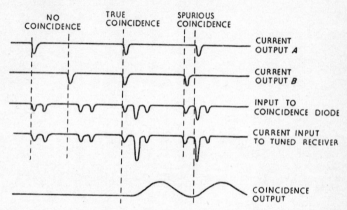

Fig. 7.32. Waveform diagram of double pulse fast coincidence circuit.

Signals from counter B are also fed into the receiver in a similar manner, so that there is again negligible final output for non-coincident pulses.

Now consider the arrival of a pair of coincident pulses from counters A and B. The difference in delay between the cables l_2 and l_1 is made equal to $T_4 - T_3$ so that the first pulse from B is superimposed upon the second pulse from A (see fig. 7.32). If these pulses were fed into the receiver in the absence of the diode there would be no resultant output, but, due to the non-linear characteristic possessed by the diode, the centre pulse of the three input pulses is more effective than the sum of the first and third pulses, so that an output is now delivered by the receiver. It should be noted that an output pulse is also obtained when a counter B pulse occurs at a time $2(T_4 - T_3)$ previous to a counter A pulse, as indicated in fig. 7.32. This represents a spurious coincidence and may lead to errors in some experiments.

The chief experimental difficulty encountered is that of equalizing the pulses in each pair fed to the diode, so that the receiver output,

APPLICATIONS TO NUCLEAR PHYSICS 263

for non-coincident pulses, shall be negligible in comparison with the coincident pulse output, even when there is a large amplitude difference (10 : 1) between the pulses from counters A and B. The coincidence resolving time is dependent upon the duration of the input pulses. The following experimental figures are given by Dicke:

$l_1 = 486$ cm $l_2 = 324$ cm

$l_3 = 320$ cm $l_4 = 160$ cm

$f_0 = 62$ Mc/s $\Delta f = 10$ Kc/s—receiver bandwidth

Resolving times of 2 mμsec. were obtained under test pulse conditions.

7.7.5.3 Double Grid Mixing Circuit

FISCHER and MARSHALL [735] have employed a 6BN6 type of valve having two control-grids operating on the one electron stream. Approximately equal control is exerted by the two grids and positive pulses, derived from the two nuclear counters, are fed to the grids separately. These are normally biassed off so that no anode current is passed unless there is a pulse present on both; the resulting anode output signal is then suitably amplified. The coincidence resolving time is again dependent upon the durations of the input pulses and on the minimum time overlap required to give a number of electrons at the anode sufficient to operate the output circuit. The amplitude of the input pulses should not be less than 3 V for satisfactory operation. The authors report that it is possible to obtain coincidence resolving times of the order of 0·3 mμsec. when the arrangement is tested on signals from a pulse generator. An electron transit-time of 2 mμsec. exists between the two control grids; this should be compensated by delaying the pulse applied to the second grid (i.e. the one nearer the anode) by a similar amount.

7.7.5.4 Cathode Ray Tube Method

The use of controlled electron beams for coincidence measurements has been little exploited; one arrangement, however, is described by HOFSTADTER and McINTYRE [736]. The pulse from counter A initiates a fast time-base sweep, whilst the pulse from counter B is applied, after a suitable delay, to the signal plates of a cathode ray tube (fig. 7.33). The screen is masked such that the B pulse will appear in the narrow aperture as shown only if the two pulses are coincident. The peak of the B pulse is allowed to be visible and the light from the screen is detected by a photomultiplier tube. An output pulse is produced by the latter only if the B pulse appears within the screen aperture; if pulses A and B are not coincident the pulse B will be displaced along the time-base, away from the aperture, and accordingly no final output is obtained.

The coincidence resolving time is determined by the duration of the B pulse and by the aperture width and the velocity of the time-base. The authors were able to obtain resolving times of 0·9 mμsec.

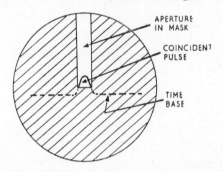

Fig. 7.33. Display of coincidence cathode ray tube.

when testing this circuit with a pulse generator. The pulse amplitude should not be less than 2·5 V and the recurrence rate of the A pulses must be sufficiently low for the triggered time-base circuit to function correctly. Recurrence frequencies of up to 1000 p.p.s. were used at resolving times of 50 mμsec.

It is worth emphasizing that the cathode ray tube method incorporates its own amplitude discriminator or pulse height selector (on the B channel), according to the manner in which the mask is cut.

Some other types of circuit have also been described by GARWIN [737], BAY and PAPP [738], and by LUNDBY [739].

We shall now consider the measurement of short time intervals involving the use of "delayed coincidence" circuits.

7.8 TIME INTERVAL MEASUREMENTS WITH DELAYED COINCIDENCE CIRCUITS

These circuits are similar to those described in § 7.7, but the variable delay line incorporated is accurately calibrated and is used for measuring the time interval.

APPLICATIONS TO NUCLEAR PHYSICS 265

The simplest application is the determination of the speed of nuclear particles by the measurement of their time of flight between two fixed points. A typical experiment is illustrated in fig. 7.34 (see reference 728). A beam of neutrons is allowed to pass through two scintillation counters A and B, separated by a suitable distance.

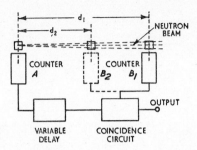

Fig. 7.34. Measurement of neutron velocity.

The coincidence output counting rate is measured for different settings of the delay line. Results are then taken again for a different distance between the two counters thus yielding the two curves shown in fig. 7.35. The curves are normalized* such that their

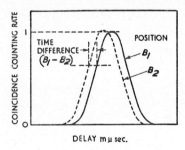

Fig. 7.35. Normalised delay coincidence curves in neutron velocity measurement.

maximum ordinates are the same; they should turn out to be of similar shape. The time delay between corresponding points on the two curves will then equal the difference in the times of flight of the neutrons between counters A and B in the two experiments. Measurement of the distances involved then yields the velocity. The

---

* The random coincidence rate should be measured and subtracted from the observed coincidence rate before the curves are normalized.

accuracy is better, the greater the change in counter spatial separation, but the maximum distance that can be employed is limited by the degree of collimation of the neutron beam; if the beam is divergent the coincidence counting rate will rapidly decrease as the distance between the counters is increased.

It is seen from fig. 7.35 that if measurements are made on the steep sides of the curves then quite small time differences may be detected by this method. The smallest discernable time difference

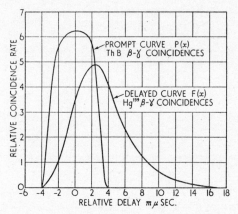

Fig. 7.36. Delayed coincidence curves in half lifetime measurement of metastable state of Hg^{199}

will only be a few percent of the coincidence resolving time; it may be remembered that the width of the curves, at half maximum amplitude points, is approximately equal to twice the coincidence resolving time.

The method has been applied to the measurement of neutron energy (since this is related to the velocity) and also to the determination of γ-ray speeds (see CLELAND and JASTRAM [740], LUCKEY and WEIL [741]). BAY estimates that time differences as short as 0·01 mμsec. may be detectable in experiments on high energy particles, using Čerenkov counters as discussed in § 7.7.4.

The second main application of the delayed coincidence technique is in the measurement of the half-life of a short lived radioactive element. A typical experiment performed by BELL provides an illustration.

When an atom of the radioactive isotope of gold Au^{199} decays, a β-particle is emitted and a mercury nucleus Hg^{199} is formed. After a short period the latter emits a γ-ray, so that, if the time interval between the β and γ emissions is measured, the lifetime of the excited state of the Hg^{199} nucleus can be determined. The γ-ray is internally converted, giving an L conversion electron, and this may be selected by a β-ray spectrometer and detected by one scintillation counter. A second counter is arranged to detect the initial β-radiation, selected by a second spectrometer, and the pulses from the two counters are fed into a delayed coincidence unit. The output coincidence rate is then measured as a function of the delay line setting, thus giving the delayed coincidence curve. After these readings have been taken the source of Au^{199} is replaced by one of ThB which emits two electrons within a time separation of less than 0·1 mμsec. A new curve of delayed coincidence rate versus delay line setting is taken, thus giving a calibration on "prompt" coincidences. The two groups of results are plotted as shown in fig. 7.36. By analysis of the curves the mean half-life of the Hg^{199} nucleus may be determined; the following method, due to NEWTON [742] (also BAY et al. [743, 744]) may be applied:

Let $P(x)$ represent the prompt coincidence curve (ThB)

$F(x)$ represent the delayed coincidence curve (Hg^{199})

and $f(t)dt$ be the probability of a delayed daughter emission (γ-ray) occurring in an interval dt at a time t after the primary emission (β-ray).

Then the delayed coincidence curve (Hg^{199}) is given by

$$F(x) = \int_{-\infty}^{\infty} f(t)P(x-t)dt \qquad (7.12)$$

For a decaying isotope we have

$$f(t) = \lambda e^{-\lambda t} \text{ for } t > 0$$

$$f(t) = 0 \qquad \text{for } t < 0$$

Hence

$$F(x) = \int_{0}^{\infty} \lambda e^{-\lambda t} P(x-t) dt \qquad (7.13)$$

which, on putting $y = x - t$, becomes

$$F(x) = \lambda e^{-\lambda x} \int_{-\infty}^{x} e^{\lambda y} P(y) dy \qquad (7.14)$$

Differentiating this relation we obtain

$$\frac{dF(x)}{dx} = \lambda [P(x) - F(x)] \qquad (7.15)$$

and

$$\frac{d \ln F(x)}{dx} = -\lambda \left[1 - \frac{P(x)}{F(x)} \right] \qquad (7.16)$$

$$= -\lambda \text{ when } F(x) \gg P(x)$$

From these equations the following properties of the curves have been deduced, provided the curves of $F(x)$ and $P(x)$ are plotted to the same included area:

(a) $F(x)$ and $P(x)$ rise at the same point on the left.

(b) $F(x)$ and $P(x)$ intersect at the maximum of $F(x)$ if only a single life-time is present.

(c) If only a single life-time is present, the curvature of $\ln F(x)$ is nowhere positive provided the curvature of $\ln P(x)$ is not positive. If this is not the case, more than one life-time is present in $F(x)$.

(d) When $F(x) \gg P(x)$ then the slope of the $\ln F(x)$ curve determines the constant λ and hence the half-life of the isotope.

(e) The centroid of $F(x)$ is displaced positively along the x axis from the centroid of $P(x)$ by an amount $1/\lambda$, the mean life-time of the delayed radiation.

Applying this analysis to the experimental results for Hg^{199} as given in fig. 7.36, BELL obtained a mean value of $2 \cdot 35 \pm 0 \cdot 2$ mμsec. for this particular γ-transition.

Half life-times of the order of $0 \cdot 1$ mμsec. have been measured by this method.

It may be emphasized that all such time difference measurements are made more easily and more accurately if the coincidence resolving time of the circuit can be lowered. It is essential to achieve very short resolving times, provided this can be done with a reasonable degree of stability, in order to reduce the number of hours required to accumulate sufficient experimental data.

7.8.1 Time Sorters.

Such units consist of multichannel delayed coincidence circuits and are designed to facilitate the plotting of the delayed coincidence curves discussed above. The pulses from the two scintillation counters are fed into a number of coincidence units through different lengths

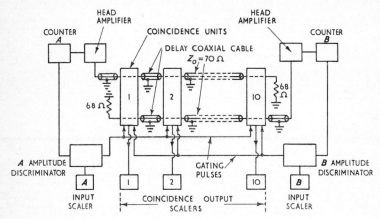

Fig. 7.37. Block schematic diagram of "timesorter" (multichannel delayed coincidence unit).

of delay cable. The output from each coincidence unit is then recorded on a separate scaler. Fig. 7.37 depicts a ten channel arrangement designed by McLusky and Moody [729]. The delay settings of the unit are adjusted to suit the experiment so that ten points on one of the curves of fig. 7.36 may be obtained for each counting run; the experimental time is accordingly reduced by a large factor. This reduction of the time required to collect the data can be of great importance in cases where the isotopes under investigation decay by an appreciable amount during the course of the experiment.

7.8.2 The Chronotron Timing Unit.

The arrangement is also similar to a multichannel delayed coincidence unit and it has been applied by Keuffel [745] and by Neddermeyer et al. [746] to the measurement of the time delay between two nuclear events, using the parallel plate counters described in § 7.4. The circuit was designed for use in a cosmic ray experiment, where the events to be measured were very infrequent, and a photographic recording technique was employed to obviate the use of scaling units.

As in the time sorter, the pulses from the two counters are fed through delay cables to ten coincidence diodes (see fig. 7.38); the delay between adjacent diodes is the same and is adjusted to suit

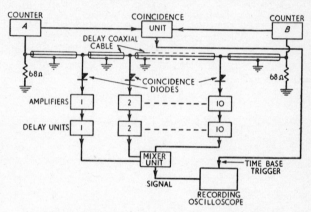

Fig. 7.38. Block schematic diagram of "Chronotron" (multi-channel time interval measuring circuit).

the experimental conditions. The arrangement thus comprises ten delayed coincidence circuits and the output from each diode is displayed on a recording cathode ray oscilloscope in the form of a pulse the amplitude of which is proportional to the diode output. The displayed pulses are separated on the oscilloscope time-base by inserting different time delays (up to several microseconds) between the output of each diode and the display mixing circuit.

The output pulses from the counters are approximately of the same amplitude (being determined by the discharge characteristic of the counters) and the pulse duration is arranged to be less than the delay between adjacent coincidence diodes, by suitable choice of values for the input differentiating capacitors and associated circuit. The diodes act as peak rectifiers, thus, if we consider a pulse from each counter travelling along the composite delay cable, the output from each of the diodes will be proportional to the larger of these two input pulses, except for the particular diode where they overlap in time. The output from the latter diode will then be proportional to the sum of the two input pulse amplitudes and hence the particular diode at which the pulses overlap in time may be at once identified from the display record (fig. 7.39). The initial time separation between the two counter discharges may then be found from a knowledge of the length of delay cables between this coincidence diode and the counters.

The oscilloscope time-base is triggered by an auxiliary coincidence circuit which gives an output signal whenever pulses from each

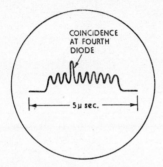

Fig. 7.39. Typical chronotron display.

counter occur within a short time interval of one another; this interval is made equal to the delay produced by the total length of cable lying between the ten coincidence diodes.

7.9 TIME INTERVAL MEASUREMENT BY INTEGRATION METHODS

We wish to measure the time interval between two pulses A and B. Let us assume that pulse B follows pulse A, then, if we allow pulse A to switch on a current I into a storage capacitor C (see fig. 7.40) and pulse B to switch this current off, the voltage developed across the storage capacitor will be dependent on the time interval between the two pulses. The capacitor is allowed to remain charged for as long a time as may be required to accurately perform the voltage amplitude measurement; it then discharges through a high resistance and returns to its original condition. If good accuracy is to be achieved it is obvious that the time which elapses between the application of the pulse A (and B) and the corresponding switching on (and off) of the current into the storage capacitor must be of known amount. The delays should also be as short as possible to avoid errors being introduced by any fluctuations in their value.

This method of time interval measurement has been applied by MOODY [747] in the millimicrosecond range. Modified forms of

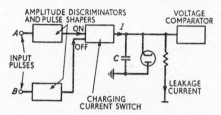

Fig. 7.40. Block schematic diagram of circuit for the measurement of time by integration.

the secondary emission valve amplitude discriminator circuit of fig. 7.11 are operated by the input pulses. Both the amplitude discriminators are followed by further trigger units, using secondary emission valves, based on the circuit of fig. 4.9. Each of the latter

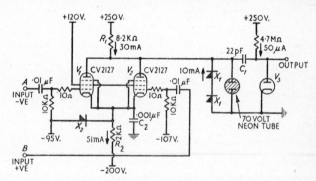

Fig. 7.41. Current integration circuit.

circuits functions primarily as an amplifier and provides a 250 mA output switching current pulse of comparatively long duration (100 μsec.) and with a leading edge rise-time of 2 to 5 mμsec. The total time delay between the leading edges of the input pulses and the subsequent output pulses is 6 to 10 mμsec. The pulse output from the A trigger circuit is arranged to be negative whilst that from the B circuit is positive; these pulses are then applied to the integrator circuit of fig. 7.41.

The valve V_1 (fig. 7.41) normally passes 51 mA of cathode current and 40 mA of anode current; valve V_2 is biassed off. A current of 30 mA flows through the resistor R_1 whilst the excess anode current in V_1 passes through the diodes X_1. When pulse A arrives, the negative pulse from the trigger circuit cuts V_1 off, so that the current through R_1 is diverted into C_1 and the stray capacitance producing a rate of rise of anode voltage of 0·75 V/mμsec. The crystal X_2 takes the current flowing through R_2 after a short time interval (C_2 is comparatively small in value). On the arrival of pulse B, a positive

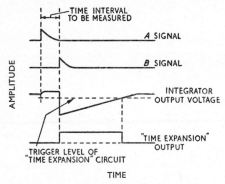

Fig. 7.42. Waveform diagram of current integration circuit.

signal is applied to valve V_2 causing it to conduct and thus returning the anode potential of V_2 (and of V_1) back towards earth. At this point in the sequence, the diode V_3 becomes open-circuited, and C_1 is left with a charge proportional to the time interval between pulses A and B. Fig. 7.42 shows the voltage waveforms across the diode V_3, the amplitudes of which may be measured; the capacitor is slowly discharged by the 50 μA leakage current.

Instead of determining the voltage change across C_1, the time taken for the capacitor to return to its normal state may be measured, using a trigger circuit. This circuit is actuated when the potential of the anode of the diode V_3 falls below a certain small value and the circuit returns to normal as soon as the diode voltage returns to a similar level. An output pulse of duration proportional to the time interval between the input A and B pulses is then obtained. The overall performance of this "time expansion" circuit is illustrated in fig. 7.43.

The above method of time interval measurement finds its most convenient application in experiments where the input pulse recurrence frequency is comparatively low, so that adequate time is available for the circuits to return to normal between pairs of input pulses. There should also be a negligible chance of a random pulse appearing within the time interval A-B since this would suppress the B pulse to be measured. If the interval A-B is very

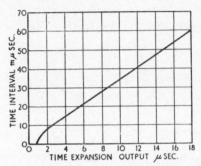

Fig. 7.43. Performance of the time expansion circuit.

small, then pulse B may be delayed by a known amount before being applied to the expander unit so that only the linear portion of the curve of fig. 7.43 is utilized in the measurement.

7.10 RECORDING OSCILLOSCOPE MEASUREMENTS

In all the previously mentioned instruments for the measurement of time intervals it has been assumed that the voltage pulses under test have fast rising leading edges free from any spurious effects. This assumption may not be justified, so that the voltage waveforms should always be examined before relying on the time interval measurements. This check is most conveniently performed by using a recording oscilloscope (see Chapter 6).

The oscilloscope itself may be employed for the measurement of time intervals and brief mention of one illustrative experiment, out of many, may be made here. POST and SHIREN [748] determined the spread of transit-time in a photomultiplier tube by recording the output current waveforms given by noise pulses, due to single electrons leaving the photocathode. A photomultiplier type 1P21 supplied with an E.H.T. voltage pulse, of 5 KV amplitude and 2 μsec. duration, at a low recurrence frequency was used and the oscilloscope time-base was triggered during each pulse. The current output pulse from the photomultiplier anode was taken through a coaxial cable to the signal plates of the cathode ray tube. The peak value of the electron multiplication of the photomultiplier was about 10^9, under these conditions, and this gain was sufficient to give an output large enough to be directly visible on the cathode ray tube screen. The noise pulses appear at random along the time-base; measurements from the film gave a mean value of 0·8 mμsec. for the duration of a pulse between half amplitude points. Sundry corrections for stray inductances and capacitances were then made, yielding a value of 0·5 mμsec. which is indicative of the spread of transit-time in the photomultiplier tube. It may be noted that the 1P21 tube cannot be run continuously with an E.H.T. of 5 KV, since sufficient feed-back occurs (presumably due to ionization) to make the tube unstable and unuseable a few microseconds after the voltage has been applied. The tube functions satisfactorily, however, provided the E.H.T. is not switched on for a period greater than 2 μsec. and is not reapplied until a comparatively long time interval has elapsed.

Particular measurements of this type, and the function of monitoring the voltage waveforms to be measured, are the chief uses of the recording cathode ray oscilloscope in the millimicrosecond region. Once the shape of the waveform is known, the circuits described in the previous Sections may be used with confidence for time interval and amplitude measurements; laborious analysis of large numbers of oscillograph film records is then obviated. When measurements are to be made on very infrequently occurring pulses, however, the oscilloscope method is valuable; the number of records is then small and the recordings may be inspected at leisure and any spurious ones rejected. This technique is used in some cosmic ray experiments where the average pulse recurrence rate is of the order of one in ten minutes or slower.

8
MISCELLANEOUS APPLICATIONS

8.1 INTRODUCTION

THE application of millimicrosecond pulse techniques to nuclear physics measurements has been described in the previous chapter and it is in this field of research that these techniques find their major use at the present time. The basic circuits, however, are not peculiar to experimental nuclear physics and can and should be applied to electronic measurements in all fields of research; it is the object of this book to bring to the notice of physicists and engineers the present day capabilities of millimicrosecond pulse circuits. A number of other applications will now be briefly described.

8.2 MILLIMICROSECOND PULSE GENERATORS FOR NARROW BANDWIDTH RADIO RECEIVER MEASUREMENTS

For a number of tests on radio and radar type narrow band receivers it is convenient to have available a signal generator giving a uniform energy output over a very wide frequency band. Measurements of the signal to noise ratio, for example, and the tracking adjustments of superheterodyne receivers can be readily made over a wide frequency range with such a generator. A pulse generator giving a very short duration output pulse is required and the output energy, over a given frequency band, can be calculated when the pulse amplitude and shape are known. Fig. 1.3 shows the frequency spectrum of a short triangular pulse. The ordinates give the relative amplitude, per unit frequency range, of the various frequency components; starting at zero frequency it is seen that these amplitudes diminish with increasing frequency. The output is reduced to $1/\sqrt{2}$ (3 db down) of the value at low frequencies when $\omega\tau/2\pi = 0.65$ that is at the frequency $650/\tau$ Mc/s where the pulse duration τ is measured in millimicroseconds.

The simplest type of pulse generator, described in § 4.2.1, is capable of giving a pulse duration of the order of 1 mμsec. corresponding to a frequency spectrum extending* up to 650 Mc/s. The repetition rate of a relay type pulse generator is rather too low for some applications and in such cases one of the valve types of pulse generator discussed in Chapter 4 may be employed. With such generators—however—the pulse duration cannot be reduced much below 4 mμsec. so that the useful output frequency spectrum scarcely exceeds 160 Mc/s. It should be noted that in these applications the output pulse must be stable in both amplitude and shape, and controllable in amplitude, if accurate receiver measurements are to be made.

8.3 USE OF MILLIMICROSECOND PULSES FOR RADAR PROPAGATION MEASUREMENTS

Very short duration pulses (3 mμsec.) have been transmitted by DELANGE [801] on microwave frequencies (4000 Mc/s) over a long path (22 miles) and the variation of the propagation time during fading periods has been investigated. Multipath transmission effects were observed with path differences of the order of 7 mμsec.

Millimicrosecond pulses are ideally suited to time measurements of this nature, where it is required to identify the pulse arriving by the direct path, out of a number of others arriving by indirect and longer paths, in the transmission medium.

8.4 THE INVESTIGATION OF ELECTRICAL DISCHARGE PHENOMENA WITH TRANSIENT RECORDING OSCILLOSCOPES

One example of the use of these oscilloscopes (see Chapter 6) is given by ENGLISH [802] in which the rate of build up of a corona discharge between point and plane electrodes was examined. The photon emission from the corona at the point electrode was detected by a photomultiplier tube and the output voltage pulse was then amplified by a distributed amplifier of 100 Mc/s bandwidth (see Chapter 5). The waveforms were recorded by an oscilloscope on a 0·2 μsec. duration time-base. The rate of photon emission from the corona discharge could then be measured from the oscilloscope photographic record.

---

* The spectrum rises again after the frequency determined by $\omega\tau/2\pi = 2$ but the amplitude is small.

8.5 ELECTRO-OPTICAL SHUTTERS FOR HIGH-SPEED PHOTOGRAPHY

There are a number of research applications which require the use of a photographic camera equipped with a shutter which can be opened for time durations lying in the millimicrosecond region. This requirement is found, for example, in the study of electrical discharges where it is desired to photograph the discharge at different instants of time during the build-up or decay process. Another application is in the study of ballistics, where it is required to photograph a projectile after a given interval of time without the film becoming fogged by the intense light flash from a previous explosion.

Mechanical shutters are too slow for operation in the millimicrosecond range and electron-optical methods are accordingly employed. Two types of electronic shutter, described below, are at present available for this purpose.

8.5.1 Kerr Cell.

These cells have been known and used for many years and ZAREM et al. [803] have give a recent description of their performance.

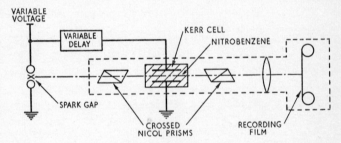

Fig. 8.1. Schematic diagram indicating use of Kerr cell camera.

Operation may be understood by considering an experiment performed by DUNNINGTON [804] for photographing the growth of a spark discharge (fig. 8.1).

The Kerr cell consists of two parallel metal plates inserted in nitrobenzene liquid contained in a glass vessel. The light from the discharge is allowed to pass between the plates as shown in the figure. Two crossed Nicol prisms are placed on either side of the cell with their respective planes of polarization making an angle of 45° with the plates of the cell. In the normal condition there is no emergent light from the system but when a potential difference is applied across the plates the nitrobenzene becomes doubly refracting and will convert the incident plane polarized wave into an elliptically polarized beam which is partially transmitted by the second Nicol prism. The delay in operation of the Kerr cell after the application of the voltage is very short and has not been measured. Similarly the cell may be shut almost instantaneously by the removal of the electric field, so that an exposure time as short as 1 mμsec. may be realized by the application of a suitable voltage pulse. Pulse amplitudes of the order of 30 KV are required.

The formula for the transmission of light through the combination of Nicol prisms and Kerr cell (neglecting absorption and reflection losses) is as follows:

$$I = I_0 \sin^2 (\pi b l E^2) \tag{8.1}$$

where I = output light intensity

I_0 = incident intensity

b = Kerr constant ($4 \cdot 1 \times 10^{-5}$ for light of wavelength 5460 Ångstroms)

l = length of plates (cm) in the direction of the optical axis

E = electric field strength in E.S.U.

In DUNNINGTON's actual experiment the cell was normally held open by a voltage applied to the spark gap. The potential difference was then raised until the spark occurred; the voltage then dropped to a low value. This voltage step was propagated down a transmission line, to provide delay, to the cell so that the voltage was removed and the shutter closed a few millimicroseconds after the spark discharge commenced. The camera recorded the spark up to the closure time of the cell and records were obtained for different lengths of transmission line. The shortest delay used by DUNNINGTON was 2 mμsec.

Recent applications of this technique have been described by ANDERSON [805] and FROOME [806] and experiments in ballistic photography are discussed by QUINN et al. [807].

An interesting application of transmission line techniques to Kerr cell design has been given by BEAMS and MORTON [808]. In their experiment the parallel plates of the cell were made very long in the direction of the light beam so that the plates formed a transmission line. The operating pulse was then applied across one end of the line and travelled along it in the same direction and with the same velocity as the incident light (requiring that the refractive index of the liquid shall be equal to the square root of the dielectric constant). Each segment of the light finds all parts of the cell in the same optical condition so that the light transmission performance of the complete shutter follows the input driving pulse shape without distortion (except for the distortion inherent in formula 8.1). A high sensitivity was obtained by this means and it was possible to use a liquid possessing a lower value of Kerr constant. This is desirable, in some cases, since the usual liquid- nitrobenzene- is chemically rather unstable and relatively opaque in the violet region. Constructional details of the cell were as follows:

Plates . . Aluminium, 335 cm long × 15·3 cm wide, separation = 4·6 mm
Line impedance . 5 Ω
Phase velocity . 1·3 × 10^{10} cm/sec.
Liquid . . Halowax Oil No. 1007

8.5.2 Image Converter.

This type of tube was first proposed by HOLST [809] whilst JENKINS and CHIPPENDALE [810] have reviewed development

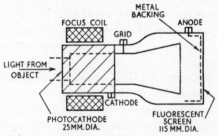

Fig. 8.2. Outline of image converter tube.
(Mullard type ME1201).

progress and described a recent design (fig. 8.2) suitable for use as a high-speed camera shutter.

An image of the object to be photographed is projected on to a photo-sensitive cathode. The resulting photo-electrons are focussed on a fluorescent screen situated at the other end of the tube by the combination of the anode to cathode electric field and the magnetic field from a focussing coil. The current in the latter is adjusted for optimum focus and a high definition display of the photocathode image may be obtained with a magnification of 4 : 1. The tube is controlled by means of a grid cylinder and the display can be switched on and off by applying suitable voltages to this grid electrode. The display is finally projected on to the photographic recording film. The fluorescent screen is metal backed to prevent any light from passing through it and fogging the film. Typical operating voltages for the tube are $+ 3$ KV on the grid and $+ 6$ KV on the anode with the cathode earthed.

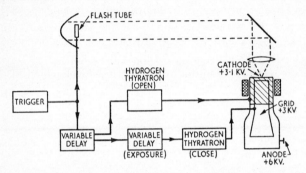

Fig. 8.3. Schematic diagram of the arrangement for pulse operation of an image converter.

A suitable pulse circuit for operating an image converter in synchronism with a light flash tube is depicted in fig. 8.3. The cathode is normally held at a potential of $+ 3\cdot1$ KV, i.e. at 100 V above the grid potential, so that no electrons can reach the anode; the shutter is then closed. The shutter is opened by triggering a hydrogen thyratron which brings the cathode potential down to the correct operating value, in the region of 60 V. An image of the photocathode picture is thus displayed on the screen and recorded on the film. When the second hydrogen thyratron is triggered, the grid potential is brought below that of the cathode and the display is removed. The time interval between the striking of the two thyratrons determines the exposure-time of the shutter; times of 50 mμsec. have been obtained in practice but considerably shorter durations should be possible.

The minimum exposure-time with both the Kerr cell and image converter types of shutter is usually limited by the amount of light available from the object to be photographed. It should be noted that the simple Kerr cell requires a pulse voltage of the order of

30 KV whilst the image converter may be operated by a pulse one tenth as great.

8.6 CONCLUSION

In experiments where the definition and measurement of short time intervals is involved, electronic pulse methods give clear, accurate and readily interpretable results, and the above examples will have sufficed to show the reader how millimicrosecond techniques may be applied. In this book the authors have endeavoured to present a concise account of present day knowledge of the subject, so that the reader may be equipped to apply such methods to his particular problems. Extension of the use of these interesting techniques will undoubtedly take place in the future and it is hoped that the volume will make some contribution to the further development of the art.

APPENDIX I

THE GENERAL DISTORTIONLESS TRANSMISSION LINE

WE wish to determine the general properties which a lossless electromagnetic wave transmission system must possess in order that an arbitrary disturbance, or signal pulse, may be propagated from one region to another without distortion of the pulse shape.

The variation with time $f(t)$ of the electric field is to be the same everywhere and may differ from point to point only in the manner of a simple delay. No *à priori* restriction is placed on the space variation of the fields.

If $\boldsymbol{E}$ is the electric field strength vector at any particular point, at time t, then we can write

$$\boldsymbol{E} = \boldsymbol{E}' \cdot f(t - T)$$

where $\boldsymbol{E}'$ is the vector amplitude function, depending solely on the coordinates, and T is a time delay, which is a scalar function of the coordinates (both $\boldsymbol{E}'$ and T will involve the coordinates of some reference point or origin). On taking the Laplace transform (see Table 1.1, page 10) we have

$$\bar{\boldsymbol{E}} = \boldsymbol{E}' \cdot e^{-pT} \cdot \bar{f} \tag{A1}$$

The magnetic field intensity $\boldsymbol{H}$ must have a similar form; we need not assume that the time variation, $g(t)$ say, and delay time, S say, are necessarily the same as for the electric field and accordingly write

$$\bar{\boldsymbol{H}} = \boldsymbol{H}' \cdot e^{-pS} \cdot \bar{g} \tag{A2}$$

Maxwell's equations for a linear lossless isotropic insulator in which there is no volume distribution of charge are

$$\nabla \times \bar{\boldsymbol{H}} = p\varepsilon\bar{\boldsymbol{E}} \tag{A3}$$

$$\nabla \times \bar{\boldsymbol{E}} = -p\mu\bar{\boldsymbol{H}} \tag{A4}$$

$$\nabla \cdot \bar{\boldsymbol{H}} = 0 \tag{A5}$$

$$\nabla \cdot \bar{\boldsymbol{E}} = 0 \tag{A6}$$

where Laplace transforms have been taken (M.K.S. units are employed).

On substituting for $\bar{E}$ and $\bar{H}$ from equations (A1) and (A2) in (A3) we find

$$[\Delta \times H' - p\nabla S \times H']e^{-pS}\bar{g} = p\varepsilon E' e^{-pT}\bar{f}$$

This must be satisfied for all values of p; thus $S = T$ and $f = g$, as fully expected; the following relations must also be satisfied

$$\nabla \times H' = 0 \qquad (A7)$$

$$\nabla T \times H' = -\varepsilon E' \qquad (A8)$$

In place of (A2) we can therefore write

$$\bar{H} = H' \cdot e^{-pT} \cdot \bar{f} \qquad (A9)$$

On substituting for $\bar{E}$ and $\bar{H}$ in (A4) we similarly find

$$\nabla \times E' = 0 \qquad (A10)$$

$$\nabla T \times E' = \mu H' \qquad (A11)$$

Further, from relations (A9) and (A5) it follows that

$$p\nabla T \cdot H' - \nabla H' = 0$$

and therefore, since this equation must be true for all values of p, we have

$$\nabla H' = 0 \qquad (A12)$$

$$\nabla T \cdot H' = 0 \qquad (A13)$$

Similarly, from relations (A1) and (A6) one obtains the following results

$$\nabla E' = 0 \qquad (A14)$$

$$\nabla T \cdot E' = 0 \qquad (A15)$$

It now remains to put a physical interpretation on the above relations.

(a) From equation (A8) and (A11) it follows that E', H' and ∇T are everywhere mutually perpendicular and form a right-handed set; these conclusions are confirmed by relations (A13) and (A15).

(b) On any surface of constant phase or constant delay time the function $f(t - T)$ is constant at any particular instant. The

APPENDIX I

electric and magnetic fields over the surface are simply proportional to E' and H' and equations (A7), (A10), (A12) and (A14) show that the fields viewed over such a surface are therefore purely electrostatic and magnetostatic. E' is thus derivable from a scalar potential, in the ordinary manner; the same applies to H'. The system must therefore embody at least two conductors in order that an electrostatic field pattern may be supported.

The line integral of E', taken over any path in the surface of constant delay, gives the potential difference between the conductors, and the discontinuity in the normal (and only) component of $\varepsilon E'$ at the surface of the conductors defines the charge density. A capacitance per unit length is then determined.

The tangential (and only) component of H' at the conductors determines the surface current, and an inductance L per unit length may be derived from the definition

$$\iiint \frac{\mu H^2}{2} d\tau = \frac{1}{2} L I^2$$

where the volume integral is taken for unit length in the appropriate direction.

(c) On substituting for H' from (A11) into (A8) we find

$$(\nabla T \cdot E')\nabla T - (\nabla T)^2 E' = -\mu\varepsilon E'$$

which, on using (A15), yields

$$|\nabla T| = +\sqrt{\mu\varepsilon} \qquad (A16)$$

Returning again to relations (A11) or (A8), and putting in this value for $|\nabla T|$, we find

$$\frac{|E'|}{|H'|} = +\sqrt{\frac{\mu}{\varepsilon}}$$

Thus, by (A1) and (A9) it is seen that $|\bar{E}|/|\bar{H}|$ and therefore $|E|/|H|$ are also equal to $\sqrt{\dfrac{\mu}{\varepsilon}}$ at any point at any time.

(d) If we move with the pulse, that is travel with a surface of constant delay, over a distance δr in a time δt we must have $\delta t - \delta T = 0$ since $f(t - T)$ is to be constant. But $\delta T = \nabla T \cdot \delta r$, thus, if δr lies along the direction of ∇T i.e. is perpendicular to the surface then

$$\frac{|\delta r|}{\delta t} = \frac{1}{|\nabla T|} = \frac{1}{\sqrt{\mu\varepsilon}} \qquad \text{by (A16)}$$

The direction of propagation is given by $\nabla T/|\nabla T|$ and the speed by $1/\sqrt{\mu\varepsilon}$. We therefore see by (a) that both the field vectors are at right angles to the direction of propagation and lie in the surface of constant delay.

(e) In the case of propagation in one dimension only, as in transmission lines consisting of parallel straight conductors, the constant delay surfaces are clearly cross-section planes. When the system consists of coaxial cones, with a common vertex but different vertical angles, we may picture a spherical wave which appears to originate from the vertex. In the intermediate case of two dimensional propagation we can imagine a cylindrical wave, confined between two parallel infinite conducting planes, spreading out radially in all directions.

APPENDIX II

TRANSMISSION LINE CHARACTERISTIC IMPEDANCES

THE majority of the following formulae have been taken from a collection made by FRANKEL [214], and are also to be found, together with results for a number of other forms, in the book by MARCHAND [254].

The conductors are supposed to be lossless and if embedded in a medium of relative permeability κ_m and relative permittivity κ_e the expressions for Z_0 (which are given in ohms) must be multiplied by $\sqrt{\kappa_m/\kappa_e}$.

The wave velocity, in the principal mode, in vacuo, is always equal to 3×10^8 m/sec. for the rectilinear types, and the inductance and capacitance are given by the relations

$$L = \frac{\kappa_m Z_0}{300} \quad \mu\text{H/m}$$

$$C = \frac{\kappa_e \times 10^4}{3 Z_0} \quad \text{pF/m}$$

APPENDIX II

Further information on various configurations is to be found in the references listed at the beginning of § 2.2.3.

OPEN LINES

Parallel strips

$377 \dfrac{a}{b}$ edge effects neglected, $a \ll b$

Parallel wires

$276 \log_{10} \left[\dfrac{d}{2a} + \sqrt{\left(\dfrac{d}{2a}\right)^2 - 1} \right]$

$\simeq 276 \log_{10} \dfrac{d}{a}$ $a \ll d$

Unequal parallel wires

$276 \log_{10} \dfrac{d}{\sqrt{ab}}$ $a \ll d$, $b \ll d$

Wire parallel to infinite plate

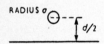

One half the values given for two parallel wires

Wire parallel to two infinite plates

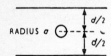

$138 \log_{10} \dfrac{2d}{\pi a}$ $a \ll d$

Wire in semi-infinite rectangular trough

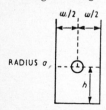

$138 \log_{10} \left[\dfrac{2w \tanh \dfrac{\pi h}{w}}{\pi a} \right]$ $a \ll h$, $w \ll h$

SCREENED LINES

Circular coaxial

$138 \log_{10} \dfrac{b}{a}$

also see graph

Circular eccentric

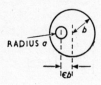

$138 \log_{10} (u - \sqrt{u^2 - 1})$

where
$$u = \frac{a^2 + b^2(1 - \varepsilon^2)}{2ab}$$

$\simeq 138 \log_{10} \dfrac{b}{a} (1 - \varepsilon^2)$

also see graph

Square outer conductor

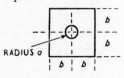

$138 \log_{10} \dfrac{1 \cdot 178 b}{a}$ $\qquad a \ll b$

Coaxial cones

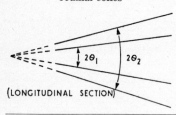

(LONGITUDINAL SECTION)

$138 \log_{10} \dfrac{\tan \theta_2/2}{\tan \theta_1/2}$

Helix with outer screen
and earthed inner rod

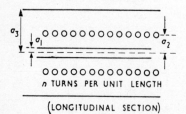

(LONGITUDINAL SECTION)

If inner rod is absent put $a_1 = 0$

If outer rod is absent put $a_3 \to \infty$

$$C = 0 \cdot 241 \times \frac{\log_{10} \dfrac{a_3}{a_1}}{\log_{10} \dfrac{a_2}{a_1} \cdot \log_{10} \dfrac{a_3}{a_2}} \text{ pF/cm}$$

$$L = 39 \cdot 5 \times 10^{-3} \cdot n^2 \frac{(a_3^2 - a_2^2)(a_2^2 - a_1^2)}{a_3^2 - a_1^2} + 4 \cdot 61 \times 10^{-3} \frac{\log_{10} \dfrac{a_2}{a_1} \cdot \log_{10} \dfrac{a_3}{a_2}}{\log_{10} \dfrac{a_3}{a_1}} \mu\text{H/cm}$$

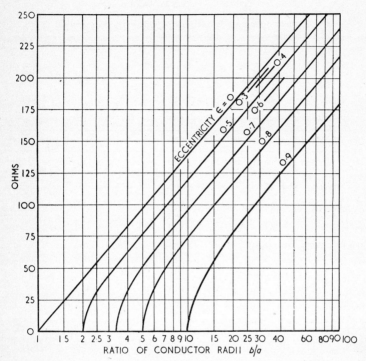

(From *Electronics*, February 1942, page 50. Copyright 1942, McGraw-Hill Publishing Company Inc.)

APPENDIX III

SOME VALVE DATA

Typical operating conditions of valves used in general pulse circuits

Type	Voltage			I_a (mA)	g_m (mA/V)	μ
	anode	screen	grid			
12AT7 CV455 Double triode	250	—	− 2	10	5·5	55
12AX7 CV492 Double triode	250	—	− 2	1·2	1·6	100
6J6 CV858 Double triode	100	—	− 0·5	9	5·6	38
EF91 CV138 H.F. pentode	250	250	− 2	10	7·5	—
6CH6 CV2127 Output pentode	250	250	− 4·5	40	11	—
EL37 — Output pentode	250	250	− 13·5	100	11	—
2D21 CV797 Thyratron	650 max.	—	—	500 peak pulse	—	—
3C45 CV372 Hydrogen thyratron	3KV max.	—	—	35A peak pulse	—	—
EFP60 — Secondary emission pentode	250	250	− 2	20	25	(Note 1)
E2133 CVX 2276 Secondary emission pentode	350	250	− 1·7	15	19	(Note 2)

Note 1: at a dynode potential of 150 V and dynode current − 16 mA.
Note 2: at a dynode potential of 250 V and dynode current − 10 mA.

APPENDIX IV

Comparative data on valves for use in linear voltage amplifiers

Type	g_m (mA/V)	C_{in} (pF) (Note 1)	C_{out} (pF) (Note 1)	working I_a (mA)	Figures of Merit	
					Economy g_m/I_a	Wideband $\dfrac{g_m \cdot 10^3}{2\pi(C_{in} + C_{out})}$ Mc/s.
6AC7 CV660	9	11	5	10	0·9	90
6AK5 CV850	5	4	3	8	0·6	120
EF91 CV138	7·5	7	2	10	0·75	130
404A	12.5	7	2·5	13	1·0	210
418A (Note 2)	25	22		50	0·5	
EFP60 anode	25	9	6	20	1·3	260
dynode	−20	9	11	−16	1·3	160
E2133 CVX2276						
anode	19	7·5	3·5	15	1·3	280
dynode	−14	7·5	3·5	−10	1·4	200
Wallmark [508]	25	3	8·5	3	8	350

Note 1: capacitance figures are quoted for the valve when cold. About 25% may be added to C_{in} when the valve is hot and both C_{in} and C_{out} may be increased by about 2 pF each, in practice, to allow for wiring strays.

Note 2: The total capacitance is reduced to 19 pF in the case of the 418A by removal of the metal shield.

APPENDIX IV

GUIDE TO NOTATION

A COMPLETE list of symbols would confuse rather than assist the reader and in any event standard symbols have been employed as far as possible. An abbreviated collection is appended here for easy reference.

ln	logarithm to the Naperian base
e	exponential function
e	electronic charge
$\log_{10}$	logarithm to the base 10

μ	magnetic permeability (M.K.S. units)
$\kappa_m = \mu/\mu_0$	permeability relative to vacuo
ε	electric pemittivity
$\kappa_e = \varepsilon/\varepsilon_0$	dielectric constant relative to vacuo
$V_x(t)$	potential difference across transmission line at the point x at the time t
T_x	delay time per unit length at the point x on a tapered line
T	delay time per unit length on a uniform line, also delay per section of a filter, also time constant of a lumped circuit
ρ_x	voltage amplitude reflection coefficient at ends of a line ($x = 0$ or l)
Z_{0x}	characteristic impedance at the point x on a tapered line
Z_0	characteristic impedance of a uniform line, also of a filter
Z_1	internal impedance of a source feeding a line
Z_2	impedance of load connected at the end of a line
—	a bar placed over a variable indicates the Laplace transform with respect to time. On putting $p = j\omega$ the same symbol denotes the complex amplitude in the case of sinusoidal oscillations
$\rightarrow, \leftarrow$	direction of wave propagation on a line
V_0	real amplitude of sinusoidally alternating voltage (compare $V_0(t)$ which represents the total voltage variation with time at the point $x = 0$ on a line)
τ	pulse duration, time interval

BIBLIOGRAPHY

BLACKBURN, F.; *Components Handbook*. M.I.T. Radiation Laboratory Series No. 7, McGraw-Hill, 1949.

BRAINERD, J. G., KOEHLER, G., REICH, H. J., WOODRUFF, L. F.; *Ultra-High-Frequency Techniques*. Van Nostrand, 1942.

CARSLAW, H. S. and JAEGER, J. C.; *Operational Methods in Applied Mathematics*. Oxford University Press, 1947.

CHANCE, B., HUGHES, V., MACNICHOL, E. F., SAYRE, D. and WILLIAMS, F. C.; *Waveforms*. M.I.T. Radiation Laboratory Series No. 19, McGraw-Hill, 1949.

CHERRY, C.; *Pulses and Transients in Communication Circuits*. Chapman & Hall, 1949.

GARDNER, M. F. and BARNES, J. L.; *Transients in Linear Systems*. Wiley, 1942.

GLASCOE, G. N. and LEBACQZ, J. V.; *Pulse Generators*. M.I.T. Radiation Laboratory Series No. 5, McGraw-Hill, 1948.

GROVER, F. W.; *Inductance Calculations*. Van Nostrand, 1946.

GUILLEMIN, E. A.; *Communication Networks*.
Vol. I, "The Classical Theory of Lumped Constant Networks".
Vol. II, "The Classical Theory of Lines, Cables and Filters". Wiley, 1931.

HERCOCK, R. J.; *The Photographic Recording of Cathode Ray Tube Traces*. Ilford Ltd., London, 1947.

JACKSON, L. C.; *Wave Filters*. Methuen, 1946.

JACKSON, WILLIS; *High Frequency Transmission Lines*. Methuen, 1945.

JACOB, L.; *An Introduction to Electron Optics*. Methuen, 1951.

KARAKASH, J. J.; *Transmission Lines and Filter Networks*. Macmillan, 1950.

MALOFF, I. G. and EPSTEIN, D. W.; *Electron Optics in Television*. McGraw-Hill, 1938.

RAMO, S. and WHINNERY, J. R.; *Fields and Waves in Modern Radio*. Wiley, 1944.

SCHELKUNOFF, S. A.; *Electromagnetic Waves*. Van Nostrand, 1943.

STARR, A. T.; *Electric Circuits and Wave Filters*. Pitman, 1946.

VALLEY, G. E., Jr. and WALLMAN, H.; *Vacuum Tube Amplifiers*. M.I.T. Radiation Laboratory Series No. 18, McGraw-Hill, 1948.

REFERENCES

Chapter 1—Theoretical Introduction

101. ELMORE, W. C.; Electronics for the nuclear physicist. *Nucleonics* 1948 **2** (Feb. 4–17, March 16–36).
102. CHENG, D. K.; A note on the reproduction of pulses. *Proc. Inst. Radio Engrs* 1952 **40** 962–5.
103. EAGLESFIELD, C. C.; Transition time and pass band. *Proc. Inst. Radio Engrs* 1947 **35** 166–7.
104. SAMULON, H. A.; Spectrum analysis of transient response curves. *Proc. Inst. Radio Engrs* 1951 **39** 175–86.
105. DI TORO, M. J.; Pulse and amplitude distortion in linear networks. *Proc. Inst. Radio Engrs* 1948 **36** 24–36.
106. TUCKER, D. G.; Bandwidth and speed of build-up as performance criteria for pulse and television amplifiers. *J. Inst. Elect. Engrs* 1947 **94** (Pt. III) 218–27.

Chapter 2—Transmission Lines

201. BOAST, W. B.; Transpositions and the calculation of inductance from geometric mean distances. *Trans. Amer. Inst. Elect. Engrs* 1950 **69** 1531–4.
202. BRAZMA, N. A.; Solution of the basic problem of electromagnetic processes in a multi-conducting system. *Dokl. Akad. Nauk SSSR* 1949 **69** 313–6.
203. BRAZMA, N. A.; A new solution of the fundamental problem of the propagation of electromagnetic waves along a set of conductors. *Dokl. Akad. Nauk SSSR* 1951 **76** 41–4.
204. BRÜDERLINK, R.; On the systematic calculation of the inductances of multiple conductor systems. *Elektrotech. Z.* (ETZ) (July 1st) 1949 **70** 233–6.
205. FRANKEL, S.; Simplified procedure for computing behaviour of multi-conductor lossless transmission lines. *Elect. Commun.* 1948 **25** 286–90.
206. KLUSS, E.; Capacitance of multiple-conductor systems. Capacitance effect on adjacent conductors. *Elektrotech. Z.* (ETZ) (Feb. 1st) 1950 **71** 63–5.
207. WAIDELICH, D. L.; Steady state waves on transmission lines. *Trans. Amer. Inst. Elect. Engrs* 1950 **69** 1521–4.
208. ANDERSON, G. M.; The calculation of the capacitance of coaxial cylinders of rectangular cross-section. *Trans. Amer. Inst. Elect. Engrs* 1950 **69** 728–31.
209. BARCLAY, W. J. and SPANGENBERG, K.; Graph of impedance of eccentric conductor cables. *Electronics* (Feb.) 1942 50.
210. BROWN, G. H.; Impedance determination of eccentric lines. *Electronics* (Feb.) 1942 49.
211. BUCHHOLZ, H.; Calculation of the wave impedance and attenuation of high-frequency transmission lines from the field of the (corresponding) perfect conductors. *Arch. Electrotech.* 1948 **39** (Sept. 79–100, Dec. 202–15).

212. CRAGGS, J. W. and TRANTER, C. J.; The capacity of twin cable. *Quart. Appl. Math.* 1945 **3** 268–72, 380–3.
213. FRANKEL, S.; Characteristic impedance of parallel wires in rectangular troughs. *Proc. Inst. Radio Engrs* 1942 **30** 182–90.
214. FRANKEL, S.; Characteristic functions of transmission lines. *Communications* (March) 1943 **23** 32–5.
215. GANS, R.; Lecher wires in a screening tube. *Publ. Fac. Cienc. Fis-Mat. La Plata* 1949 (No. 196) 223–50.
216. LANSBERG, M.; A minimum problem as basis for calculating capacitances of cables. *Z. angew Math. Mech.* (May) 1948 **28** 143–52.
217. MEINKE, H. H.; The behaviour at high frequencies of curved homogeneous lines. *Arch. elekt. Übertragung* (March) 1951 **5** 106–12.
218. PARZEN, J. P.; The capacity per unit length and characteristic impedance of coaxial cables with one slightly non-circular conductor. *J. Appl. Phys.* 1947 **18** 774–9.
219. RE QUA, F. L.; Resistance and capacitance relations between short cylindrical conductors. *Trans. Amer. Inst. Elect. Engrs* 1945 **64** 724–30.
220. ROTHE, P. G.; Approximate formulae for the characteristic impedances of some conical transmission lines. *R.A.E. Tech. Note* RAD 532, January 1953.
221. SHEBES, M. R.; Coaxial cable with confocal cross-section of cylinder type. *Radiotekhnika* (July–August) 1949 **4** 36–44.
222. TSEITLIN, L. A.; The determination of the average potential and capacitance of linear conductors. *J. Tech. Phys. USSR* 1946 **16** (No. 1) 123–7.
223. TSEITLIN, L. A.; Parameters of systems of rectilinear and curvilinear conductors. *Elektrichestvo* 1948 (No. 4) 31–6.
224. TSEITLIN, L. A.; Capacitance of curvilinear conductors. *Dokl. Akad. Nauk SSSR* 1948 **59** 1583–6.
225. WHEELER, H. A.; Transmission line impedance curves. *Proc. Inst. Radio Engrs* 1950 **38** 1400–3.
226. WISE, W. H.; Capacity of a pair of insulated wires. *Quart. Appl. Math.* 1949 **7** 432–6.
227. KENNEY, N. D.; Coaxial cable design. *Electronics* (May) 1945 **18** 124–8.
228. SMITH, E. W.; Radar cables. *Wireless World* (April) 1946 **52** 129–31.
229. STANFORD, N. C. and QUARMBY, R. B.; Characteristics of r.f. cables. *Wireless Engr* 1946 **23** 295–8.
230. ZIMMERMAN, K. H.; High-frequency cable design. *Electronics* (Feb.) 1948 **21** 112–15.
231. Reports of conference on radio-frequency cables. *Trans. Amer. Inst. Elect. Engrs* **64**, Supplements, Dec. 1945.
232. BONDI, H. and KUHN, S.; Concentric line. Critical wavelength of the higher modes. *Wireless Engr* 1947 **24** 222–3.
233. WHINNERY, J. R. and JAMIESON, H. W.; Equivalent circuits for discontinuities in transmission lines. *Proc. Inst. Radio Engrs* 1944 **32** 98–114.
234. WHINNERY, J. R., JAMIESON, H. W. and ROBBINS, T. E.; Coaxial line discontinuities. *Proc. Inst Radio Engrs* 1944 **32** 695–709.
235. MILES, J. W.; Plane discontinuities in coaxial lines. *Proc. Inst. Radio Engrs* 1947 **35** 1498–1502.
236. MARCUS, P. M.; The interaction of discontinuities on a transmission line. *M.I.T. Radiation Laboratory Report* No. 930, Feb. 1946.
237. CORNES, R. W.; A coaxial line support for 0–4000 Mc/s. *Proc. Inst. Radio Engrs* 1949 **37** 94–7.

238. KADEN, H. and ELLENBERGER, G.; Reflection-free support discs in coaxial lines. *Arch. elect. Übertragung* 1949 **3** 313–22.
239. PETERSON, D. W.; Note on a coaxial line bead. *Proc. Inst. Radio Engrs* 1949 **37** 1294.
240. ROWLAND, H. J.; The series reactance on coaxial lines. *Proc. Inst. Radio Engrs* 1948 **36** 65–9.
241. GARWIN, R. L.; A pulse generator for the millimicrosecond range. *Rev. Sci. Instrum.* 1950 **21** 903–4.
242. CROSBY, D. R. and PENNYPACKER, C. H.; Radio-frequency resistors as uniform transmission lines. *Proc. Inst. Radio Engrs* 1946 **34** 62–6P
243. CLEMENS, G. J.; A tapered line termination at microwaves. *Quart. Appl. Math.* 1949 **7** 425–32.
244. COUANAULT, G. and HERRENG, P.; Study of impedance irregularities of coaxial cables by oscillographic observation of pulse echoes. *Câbles et Transm.* (July) 1948 **2** 219–32.
245. COX, C. R.; Design data for beaded coaxial lines. *Electronics* (May) 1946 **19** 130–5.
246. FUCHS, G.; Reflections in a coaxial cable due to impedance irregularities. *Instn Elect. Engrs Monograph* No. 25, February 1952; *Proc. Instn Elect. Engrs* (April) 1952, Part IV.
247. HERRENG, P. and VILLE, J.; Study of impedance irregularities of coaxial cables by oscillographic observation of pulse echoes. *Câbles et Transm.* (April) 1948 **2** 111–30.
248. RAYMOND, F.; Note on the propagation of an electromagnetic signal along a heterogeneous line. *Comptes Rend. Acad. Sci. Paris* 1946 **222** 1000–2 also 1945 **220** 450.
249. SAPIR, I. L.; Calculation of transient processes in composite lines with distributed and lumped constants. *Elektrichestvo* 1951 (No. 3) 28–32.
250. KING, R. and TOMIYASU, K.; Terminal impedance and generalized two-wire line theory. *Proc. Inst. Radio Engrs* 1949 **37** 1134–9.
251. TOMIYASU, K.; The effect of a bend and other discontinuities on a two-wire transmission line. *Proc. Inst. Radio Engrs* 1950 **38** 679–82.
252. OLIVER, M. H.; Discontinuities in concentric line impedance measuring apparatus. *J. Instn Elect. Engrs* 1950 **97** (Pt. III) 29.
253. MARCHAND, N.; Complex transmission line network analysis. *Elect. Communication* 1944 **22** 124–9.
254. MARCHAND, N.; Ultra-high frequency transmission and radiation. Wiley & Sons, 1947, p. 271–9.
255. MILLER, K. W.; Diffusion of electric currents into rods, tubes and flat surfaces. *Trans. Amer. Inst. Elect. Engrs* 1947 **66** 1496–1502.
256. SIM, A. C.; The pulse resistance of circular hollow wires. *B.T.H. Research Laboratory Report* No. L3846, 1949.
257. SIM, A. C.; A theory of the transient skin effect. *B.T.H. Research Laboratory Report* No. L3901, 1949.
258. VALLESE, L. M.; Density distribution of transient currents in conductors. *Proc. Inst. Radio Engrs* 1950 **38** 563.
259. WHINNERY, J. R.; Skin effect formulae. *Electronics* (February) 1942 **15** 44–8.
260. SMITH, P. H.; Optimum coax diameters. *Electronics* (February) 1950 **23** 111–4.
261. MORENO, T.; Microwave transmission design data. McGraw-Hill 1948, p. 201–9.

262. WILLIAMS, E. M. and SCHATZ, E. R.; Design of exponential line pulse transformers. *Proc. Inst. Radio Engrs* 1951 **39** 84–6.
263. BRECKINRIDGE, R. G. and THURNAUER, H.; Digest of the literature on dielectrics. *Publ. Nat. Res. Coun. Nat. Acad. Sci. (Wash.)* (Sept.) 1951, No. 202.
264. CLOGSTON, A. M.; Reduction of skin effect losses by the use of laminated conductors. *Proc. Inst. Radio Engrs* 1951 **39** 767–82.
265. BLACK, H. S., MALLINCKRODT, C. O. and MORGAN, S. P.; Experimental verification of the theory of laminated conductors. *Proc. Inst. Radio Engrs* 1952 **40** 902–5.
266. WINKLER, M. R. and KALLMANN, H. E.; Discussion on reference 282 below. *Proc. Inst. Radio Engrs* 1947 **35** 1097–100.
267. FRANKEL, S.; Contribution to previous discussion. *Proc. Inst. Radio Engrs* 1949 **37** 406.
268. OGLAND, J. W.: Transmission line used as a pulse transformer. *R.R.D.E. Report* No. 284, 1945.
269. PALMERO, A. J.; Distributed capacity of single layer coils. *Proc. Inst. Radio Engrs* 1934 **22** 897–905.
270. BLEWETT, J. P., LANGMUIR, R. V., NELSON, R. B. and RUBEL, J. H.; Delay lines. *G.E.C. Report*, May 1943.
271. BLEWETT, J. P. and RUBEL, J. H.; Video delay lines. *Proc. Inst. Radio Engrs* 1947 **35** 1580–4.
272. LOSHAKOV, L. N. and OLDEROGGE, E. B.; Theory of coaxial spiral line. *Radiotekhnika* 1948 **3** (No. 2) 11–20.
273. LEWIS, I. A. D.; Note on the variations of phase velocity in continuously-wound delay lines. *Proc. Instn Elect. Engrs* 1951 **98** (Pt. III) 312–4, also *Proc. Instn Elect. Engrs* 1952 **99** (Pt. III) 158.
274. KALLMANN, H. E.; Equalized delay lines. *Proc. Inst. Radio Engrs* 1946 **34** 646–57.
275. SHAW, E. A. G.; A theoretical study of the delay cable and the concentric cable with a metal foil outer conductor. *M.A.P.H.Q.* 1943 *C.R.B.* Ref. 47/915.
276. ERICKSON, R. A. and SOMMER, H.; The compensation of delay distortion in video delay lines. *Proc. Inst. Radio Engrs* 1950 **38** 1036–40.
277. WEEKES, D. F.; A video delay line. *M.I.T. Radiation Laboratory Report* No. 302, 1943.
278. DI TORO, J. J.; Phase corrected delay lines. *Inst. Radio Engrs National Convention*, March 1948.
279. LUND, C. O.; A broadband transition from coaxial line to helix. *R.C.A. Review* (March) 1950 **11** 133–42.
280. ESSEN, L.; The propagation constants of delay cables. *M.A.P./N.P.L. Report* No. 16, 1943.
281. HODELIN, J. A.; Coaxial cable with high characteristic impedance. *Radio Franc.* (Feb.) 1949 23–4.
282. KALLMANN, H. E.; High impedance cable. *Proc. Inst. Radio Engrs* 1946 **34** 348–51 (Wav. and El.).
283. RUBEL, J. H., STEVENS, H. E. and TROELL, R. E.; Design of delay lines. *G.E.C. Report*, Oct. 1943.
284. ZIMMERMAN, K. H.; Spiral delay lines. *Elec. Communication* 1946 **23** 327–8.
285. HEBB, M. H., HORTON, C. W. and JONES, F. B.; On the design of networks for constant time delay. *J. Appl. Phys.* 1949 **20** 616–20.
286. KALLMANN, H. E., SPENCER, R. E. and SINGER, C. P.; Transient response. *Proc. Inst. Radio Engrs* 1945 **33** 169–95.

287. THOMSON, W. E.; Delay networks having maximally flat frequency characteristics. *Proc. Instn Elect. Engrs* 1949 **96** (Pt. III) 487–91.
288. TREVOR, J. B.; Artificial delay line design. *Electronics* (June) 1945 135–7.
289. WHEELER, H. A. and MURNAGHAN, F. D.; Theory of wave filters containing a finite number of sections. *Phil. Mag.* (July) 1928 **6** 146–74.
290. DAWES, C. L., THOMAS, C. H. and DROUGHT, A. B.; Impulse measurements by repeated structure networks. *Trans. Amer. Inst. Elect. Engrs* 1950 **69** (Pt. I) 571–83.
291. GIACOLETTO, L. J.; Optimum resistive terminations for single section constant-k ladder type filters. *R.C.A. Rev.* (Sept.) 1947 **8** 460–79.
292. FERGUSON, A. J.; A note on phase correction in electrical delay networks. *Canad. J. Res.* (Jan.) 1947 A**25** 68–71.
293. GOLAY, M. J. E.; The ideal low-pass filter in the form of a dispersionless lag line. *Proc. Inst. Radio Engrs* 1946 **34** 138–44P (Wav. and El.).

Chapter 3—Transformers

301. FANO, R. M.; Theoretical limitations to the broadband matching of arbitrary impedances. *J. Franklin Inst.* 1950 **249** 57–84 and 139–54.
302. MARSHALL, J.; Transmission line reflection doubling amplifier. *Rev. Sci. Instrum.* 1950 **21** 1010–3.
303. HADLOCK, C. K.; Some studies of pulse transformer equivalent circuits. *Proc. Inst. Radio Engrs* 1951 **39** 81–3.
304. MAURICE, D. and MINNS, R. H.; Very wide-band radio-frequency transformers. *Wireless Engr.* 1947 **24** 168–77 and 209–16.
305. MELVILLE, W. S.; Theory and design of high-power pulse transformers. *J. Instn Elect. Engrs* 1946 **93** (Pt. IIIA) 1063–80.
306. MOODY, N. F.; A treatise on the design of pulse transformers for handling small powers. *T.R.E. Technical Monograph* No. 5A, 1943.
307. MOODY, N. F.; Low power pulse transformers. *J. Instn Elect. Engrs* 1946 **93** (Pt. IIIA) 311–2.
308. MOODY, N. F., MCLUSKY, C. J. R. and DEIGHTON, M. O.; Millimicrosecond Pulse techniques. *Electronic Engng* 1952 **24**.
   pp. 214–9, Part I—An introduction to techniques and the development of basic circuits.
   pp. 289–94, Part III—Time expansion for millimicrosecond pulse intervals.
309. RUDENBERG, R.; Electric oscillations and surges in subdivided windings. *J. Appl. Phys.* 1940 **11** 665–80.
310. RUDENBERG, R.; Surge characteristics of two-winding transformers. *Trans. Amer. Inst. Elect. Engrs* 1941 **60** 1136–44.
311. RUDENBERG, R.; Electromagnetic waves in transformer coils treated by Maxwell's equations. *J. Appl. Phys.* 1941 **12** 219–29.
312. SLATER, J. C.; Microwave transmission. McGraw-Hill, 1942, p. 69–75.
313. FRANK, N. H.; Reflections from sections of tapered transmission lines and waveguides. *M.I.T. Radiation Laboratory Report* No. 17 (Sec. 43), January 1943.
314. PIERCE, J. R.; Note on the transmission line equations in terms of impedance. *Bell Syst. Tech. J.* 1943 **22** 263–5.
315. WALKER, L. R. and WAX, N.; Non-uniform transmission lines and reflection coefficients. *J. Appl. Phys.* 1946 **17** 1043–5.
316. BOLINDER, F.; Fourier transforms in the theory of inhomogeneous transmission lines. *Trans. Royal Inst. Tech.* (Stockholm) 1951, No. 48.

317. CARSON, J. R.; Propagation of periodic currents over non-uniform lines. *Electrician* (March) 1921 **86** 272–3.
318. ILIN, V. A.; Long lines with parameters varying along the line. *Elektrichestvo* 1950 (No. 2) 54–9.
319. RAYMOND, F. H.; Propagation in a non-homogeneous line. *J. Phys. Radium* (Sér. 8) June 1946 **7** 171–7.
320. ZIN, G.; Equations of incident and reflected waves in non-uniform lines under stationary conditions. *Alta Frequenza* 1941 **10** 149–78.
321. ZIN, G.; Transients in non-uniform lines. *Alta Frequenza* 1941 **10** 707–50.
322. BURROWS, C. R.; The exponential transmission line. *Bell Syst. Tech. J.* 1938 **17** 555–73. Exponential transmission line. *Communications* 1938 **18** October 7–9, 26–8 and November 11–3, 17–8.
323. WHEELER, H. A.; Transmission lines with exponential taper. *Proc. Inst. Radio Engrs* 1939 **27** 65–71.
324. RUHRMANN, A.; Improvements of the transformation properties of exponential lines by compensation arrangements. *Arch. elekt. Übertragung* 1950 **4** 23–32.
325. MILNOR, J. W.; The tapered transmission line. *Trans. Amer. Inst. Elect. Engrs* 1945 **64** 345–6.
326. WHEELER, H. A. and MURNAGHAN, F. D.; Theory of wave filters containing a finite number of sections. *Phil. Mag.* (July) 1928 **6** 146–174.
327. OGLAND, J. W.; Transmission line used as a pulse transformer. *R.R.D.E. Report* No. 284, 1945.
328. SCHATZ, E. R. and WILLIAMS, E. M.; Pulse transients in exponential transmission lines. *Proc. Inst. Radio Engrs* 1950 **38** 1208–12.
329. WILLIAMS, E. M. and SCHATZ, E. R.; Design of exponential line pulse transformers. *Proc. Inst. Radio Engrs* 1951 **39** 84–6.
330. GENT, A. W. and WALLIS, P. J.; Impedance matching by tapered transmission lines. *J. Instn Elect. Engrs* 1946 **93** (Pt. IIIA) 559–63.
331. BALLANTINE, S.; Non-uniform lumped electric lines. *J. Franklin Inst.* 1927 **203** 561–82.
332. STARR, A. T.; The non-uniform transmission line. *Proc. Inst. Radio Engrs* 1932 **20** 1052–63.
333. ARNOLD, J. W. and BECHBERGER, P. F.; Sinusoidal currents in linearly tapered loaded transmission lines. *Proc. Inst. Radio Engrs* 1931 **19** 304–10.
334. ARNOLD, J. W. and TAYLOR, R. C.; Linearly tapered loaded transmission lines. *Proc. Inst. Radio Engrs* 1932 **20** 1811–7.
335. CHRISTIANSEN, W. N.; An exponential transmission line employing straight conductors. *Proc. Inst. Radio Engrs* 1947 **35** 576–81 (Wav. and El.).
336. BROWN, A. H.; Networks for balanced to unbalanced impedance transformation. *R.A.E. Tech. Note* No. Rad 427, 1948.
337. MARCHAND, N.; Transmission line conversion transformers. *Electronics* (Dec.) 1944 **17** 142–5.
338. MARCHAND, N.; Ultra-high frequency transmission and radiation. Wiley & Sons 1947, p. 279–85.
339. FUBINI, E. G. and SUTRO, P. S.; A wide-band transformer from an unbalanced to a balanced line. *Proc. Inst. Radio Engrs* 1947 **35** 1153–5.
340. LEWIS, I. A. D. and WHITBY, H. C.; A simple inverting transformer for millimicrosecond pulses. *A.E.R.E. Report* G/R 751, 1951. Pat. Applcn No. 29008/51: Improvements in or relating to electrical phase inversion systems.

341. ROCHELLE, R. W.; Transmission line pulse inverter. *Rev. Sci. Instrum.* 1952 **23** 298–300.
342. FUCHS, M.; Intercoupled transmission lines at radio frequencies. *Elect. Communication* 1944 **21** 248–56.
343. BLOCH, A.; Mutual capacitance and inductance of parallel lines. *Wireless Engr* 1944 **21** 280–1.
344. KARAKASH, J. J. and MODE, D. E.; A coupled coaxial transmission line band-pass filter. *Proc. Inst. Radio Engrs* 1950 **38** 48–52.
345. CROUT, P. D.; A method of virtual displacements for electrical systems with applications to pulse transformers. *Proc. Inst. Radio Engrs* 1946 **35** 1236–47.
346. LEWIS, I. A. D.; Analysis of a helical transmission line with thermionic valve action. *A.E.R.E. Report* G/R 611 January 1950. Analysis of a transmission-line type of thermionic amplifier valve. *Instn Elect. Engrs Monograph* No. 57, Dec., 1952. *Proc. Instn Elect. Engrs* 1953, Pt. IV.
347. MARSON, A. E. and ADCOCK, M. D.; Radiation from helices. *Overseas Research Report* D.S.I.R. 209/51. Off. Tech. Serv. Washington PB. 101175.
348. MATHEWS, W. E.; Travelling-wave amplification by means of coupled transmission lines. *Proc. Inst. Radio Engrs* 1951 **39** 1044–51.
349. HUMPHREYS, B. L., KITE, L. V. and JAMES, E. G.; Phase velocity of waves in a double helix. *G.E.C. Report* No. 9507, 1948.
350. RUDENBERG, H. G.; Pulse transformer considered as a wide-band network. *Proc. Inst. Radio Engrs* 1951 **39** 306.
351. RUDENBERG, H. G.; The distributed transformer. Research Division, Raytheon Manufacturing Co., April 1952.

Chapter 4—Pulse Generators

401. WELLS, F. H.; Millimicrosecond pulse circuits. *J. Brit. Instn Radio Engrs* 1951 **11** 491–503.
402. ESPLEY, D. C.; Generation of very short pulses. *J. Instn Elect. Engrs* 1946 **93** (Pt. IIIA) 314–5.
403. ESPLEY, D. C., CHERRY, E. C. and LEVY, M. M.; The pulse testing of wide-band networks. *J. Instn Elect. Engrs* 1946 **93** (Pt. IIIA) 1176–87.
404. WHITE, E. L. C.; The use of delay networks in pulse formation. *J. Instn Elect. Engrs* 1946 **93** (Pt. IIIA) 312–4.
405. BROWN, J. T. L. and POLLARD, C. E.; Mercury contact relays. *Electrical Engng* 1947 **66** 1106–9.
406. MOODY, N. F., MCLUSKY, G. J. R. and DEIGHTON, M. O.; Millimicrosecond pulse techniques. *Electronic Engng* 1952 **24**, p. 214–9; Part I— An introduction to techniques and the development of basic circuits.
407. GARWIN, R. L.; A pulse generator for the millimicrosecond range. *Rev. Sci. Instrum.* 1950 **21** 903–4.
408. WHITBY, H. C.; Improvements in or relating to switch mechanisms. Prov. Pat. Applicn, Nos. 3502/52 and 11256/52.
409. FLETCHER, R. C.; Impulse breakdown in the 10^{-9} sec. range of air at atmospheric pressure. *M.I.T. Laboratory for Insulation Research, Tech. Reports* Nos. 20 and 21, 1949.
410. GOODMAN, D. H., SLOAN, D. H. and TRAU, E.; A high voltage high speed square wave surge generator. *Rev. Sci. Instrum.* 1952 **23** 766–7.
411. IVES, R. L.; The relay oscillator and related devices. *J. Franklin Inst.* 1946 **242** 243–77.

412. McALISTER, K. R.; A variable length radio-frequency transmission line section. *J. Sci. Instrum.* 1951 **28** 142–3.
413. BROWN, D. R.; An interim report on tests to study the long term stability of braided r.f. cables. *R.A.E. Tech. Note* No. RAD 476, 1950.
414. FINCH, T. R.; An impulse generator electronic switch for visual testing of wide-band networks. *Proc. Inst. Radio Engrs* 1950 **38** 657–61.
415. BIRNBAUM, M.; A method for the measurement of ionization and deionization times of thyratron tubes. *Trans. Amer. Inst. Elect. Engrs.* 1948 **67** 209–14.
416. CHANCE, J. C. R.; Note on the ionization and de-ionization times of gas-filled thyratrons. *J. Sci. Instrum.* 1946 **23** 50–2.
417. KNIGHT, H. DE B. and HERBERT, L.; The development of mercury vapour thyratrons for radar modulator service. *J. Instn Elect. Engrs* 1946 **93** (Pt. IIIA) 949–62.
418. KNIGHT, H. DE B.; Hot cathode thyratrons: practical studies of characteristics. *Proc. Instn Elect. Engrs* 1949 **96** 361–378, 379–381.
419. MULLIN, C. J.; Initiation of discharge in arcs of thyratron type. *Phys. Rev.* 1946 **70** 401–5.
420. ROMANOWITZ, H. A. and DOW, W. G.; Statistical nature and physical concepts of thyratron de-ionization time. *Trans. Amer. Inst. Elect. Engrs* 1950 **69** 368–79.
421. WEBSTER, E. W.; Note on the ionization time of an argon filled relay. *J. Sci. Instrum.* 1947 **24** 299–301.
422. WITTENBERG, H. H.; Thyratrons in radar modulator service. *R.C.A. Rev.* (March) 1949 **10** 116–33.
423. YU, Y. P., CHRISTALDI, P. S. and KALLMANN, H. E.; Progress report on millimicrosecond oscillography. *Proc. Nat. Electronics Conf.* 1950 **6** 360–72.
424. FOWLER, C. S.; A narrow pulse generator. *Wireless Engr* 1950 **27** 265–9.
425. KNIGHT, H. DE B. and HOOKER, O. N.; The hot cathode hydrogen filled thyratron. *B.T.H. Research Laboratory Report* No. RLP 203.
426. HEINS, H.; The hydrogen thyratron. *Instruments* (April) 1946 **19** 211 and 250.
427. WOODFORD, J. B. and WILLIAMS, E. M.; The initial conduction interval in high-speed thyratrons. *J. Appl. Phys.* 1952 **23** 722–4.
428. HEALEA, M.; Bibliography; Counting circuits and secondary emission. *Nucleonics* (March) 1948 **2** 66–74.
429. ALLEN, J. S.; Recent applications of electron multiplier tubes. *Proc. Inst. Radio Engrs* 1950 **38** 346–58.
430. KEEP, D. N.; Wide-band logarithmic amplifier using secondary emission valves. *A.S.R.E. Tech. Note* TX/50/9 1950.
431. MEULLER, C. W.; Receiving tubes employing secondary electron emitting surfaces exposed to the evaporation from oxide cathodes. *Proc. Inst. Radio Engrs* 1950 **38** 159–64.
432. OVERBEEK, A. J. W. M. VAN; Voltage controlled secondary emission multipliers. *Wireless Engr* 1951 **8** 114–25.
433. DIEMER, G. and JONKER, J. L. H.; Secondary-emission valve. Wide-band amplification for decimetre waves. *Wireless Engr* 1950 **27** 137–42.
434. WELLS, F. H.; Fast pulse circuit techniques for scintillation counters. *Nucleonics* (April) 1952 **10** 28–33.
435. HASTED, J. B.; Millimicrosecond pulse generation by electron bunching. *Proc. Phys. Soc.* 1948 **60** 397.
436. ADLER, R.; The 6BN6 gated beam tube. Part I—The laboratory

prototype and its circuit applications. *Proc. Nat. Electronics Conf.* 1949 **5** 408–16.
437. HAASE, A. P.; The 6BN6 gated beam tube. Part II—The commercial realization of the 6BN6. *Proc. Nat. Electronic Conf.* 1949 **5** 417–26.
438. MILLER, C. F. and MACLEAN, W. S.; New design for a secondary-emission trigger tube. *Proc. Inst. Radio Engrs* 1949 **37** 952–4.
439. BENJAMIN, R.; Blocking oscillators. *J. Instn Elect. Engrs* 1946 **93** (Pt. IIIA) 1159–75.
440. O'DELL, D. T.; A reflex peak voltmeter for very short pulses. *A.S.R.E. Tech. Note* GX/51/2, 1951.
441. HUSSEY, L. W.; Non-linear coil generators of short pulses. *Proc. Inst. Radio Engrs* 1950 **38** 40–4.
442. MELVILLE, W. S.; The use of saturable reactors as discharge devices for pulse generators. *Proc. Instn Elect. Engrs* 1951 **98** (Pt. III) 185–207.
443. BOELLA, M.; Behaviour of high resistances at radio frequencies. *Alta Frequenza* 1934 **3** 132–48.
444. HOWE, G. W. O.; Behaviour of high resistances at high frequencies. *Wireless Engr* 1935 **12** 291–5.
445. HOWE, G. W. O.; Behaviour of resistors at high frequencies. *Wireless Engr* 1940 **17** 471–7.
446. PAVLASEK, T. J. F. and HOWES, F. S.; R. F. characteristics of JAN-R-11 Type fixed composition resistors. McGill University, Dept. Elect. Engrg, 1948; Defence Research Board Canada Report D.R.9.
447. PUCKLE, O. S.; Behaviour of high resistances at high frequencies: Boella effect. *Wireless Engr* 1935 **12** 303–9.
448. SIMMONDS, J. C.; Apparatus for measurements on balanced-pair h.f. cables in the range 10–200 Mc/s. *J. Instn Elect. Engrs* 1945 **92** (Pt. III) 282–6.
449. PAVLASEK, T. J. F. and HOWES, F. S.; The effect of frequency characteristics of JAN–R–11 composition resistors on power dissipated under pulsed conditions. McGill University, Dept. Elect, Engng, 1948; Defence Research Board Canada Report D.R. 10.
450. PLANER, G. V. and PLANER, F. E.; High stability carbon resistors. *Electronic Engng* 1946 **18** 66–8, 97.
451. GRISDALE, R. O., PFISTER, A. C. and TEAL, G. K.; Borocarbon film resistors. *Proc. Nat. Electronic Conf.* 1950 **6** 441–2.
452. HERITAGE, R. J.; Metal film resistors. *Electronic Engng* 1952 **24** 324–7.
453. DAWES, C. L., THOMAS, C. H. and DROUGHT, A. B.; Impulse measurements by repeated structure networks. *Trans. Amer. Inst. Engrs* 1950 **69** 571–83.
454. ROOSBROCK, W. VAN; H.F. deposited carbon resistors. *Bell Laboratories Record* 1948 **26** 407–11.
455. ELLIOTT, J. S.; Coaxial attenuation standards. *Bell Laboratories Record* 1949 **27** 221–4.
456. KOHN, C.; The design of an r.f. coaxial resistor. *S.R.D.E. Tech. Memo.* No. 94 1948.
457. BURKHARDTSMAIER, W.; The production of pure ohmic resistances independent of frequency for short waves. *Funk u. Ton* (August) 1948 381–91.
458. CARLIN, H. J.; Broad-band dissipative matching structures for microwaves. *Proc. Inst. Radio Engrs* 1949 **37** 644–50.
459. SELBY, M. C.; High frequency voltage measurements. *Nat. Bureau of Standards, Washington, Report* CRPL-8-2 1948.

Chapter 5—Amplifiers

501. EMMS, E. T.; A critical survey of high-slope pentodes for television and f.m. reception. *Mullard Research Lab.* 1949 Report No. 34.
502. FERRIS, W. R.; Input resistance of vacuum tubes as u.h.f. amplifiers. *Proc. Inst. Radio Engrs* 1936 **24** 82–107.
503. MOODY, N. F., MCLUSKY, G. J. R. and DEIGHTON, M. O.; Millimicrosecond pulse techniques. *Electronic Engng* 1952 **24** 214–9. Part I—An introduction to techniques and the development of basic circuits.
504. DEIGHTON, M. O.; Note on a resonance effect in Hi-K ceramic disc condensers. *N.R.C. Canada, Report* ELI-2.
505. FORD, G. T.; The 404A, a broadband amplifier tube. *Bell Laboratories Record* 1949 **27** 59–61.
506. FORD, G. T. and WALSH, E. J.; The development of electron tubes for a new coaxial transmission system. *Bell Syst. Tech. J.* 1951 **30** 1103–28.
507. WALSH, E. J.; Fine wire type vacuum tube grid. *Bell Laboratories Record* 1950 **28** 165–7.
508. WALLMARK, J. T.; An experimental high-transconductance tube using space charge deflection of the electron beam. *Proc. Inst. Radio Engrs* 1952 **40** 41–8.
509. HANSEN, W. W.; On the maximum gain-bandwidth product in amplifiers. *J. Appl. Phys.* 1945 **16** 528–34.
510. HEROLD, E. W.; H.F. correction in resistance coupled amplifiers. *Communications* (August) 1938 **18** 11–4, 22.
511. SEAL, P. M.; Square wave analysis of compensated amplifiers. *Proc. Inst. Radio Engrs* 1949 **37** 48–58.
512. BEDFORD, A. V. and FREDENDALL, G. L.; Transient response of multistage video-frequency amplifiers. *Proc. Inst. Radio Engrs* 1939 **27** 277–85.
513. GIACOLETTO, L. J.; Optimum resistive terminations for single section constant-k ladder type filters. *R.C.A. Rev.* (Sept.) 1947 **8** 460–79.
514. BORG, H.; A note on wide-band amplifiers covering a frequency range up to 150 Mc/s. *A.S.R.E.* 1950 Tech. Note R4/50/18.
515. PERCIVAL, W. S.; Improvements in and relating to electron discharge devices. Pat. Spec. No. 464977, 1937.
516. ESPLEY, D. C.; Generation of very short pulses. *J. Instn. Elect. Engrs* 1946 **93** (Pt. IIIA) 314–5.
517. GINZTON, E. L., HEWLETT, W. R., JASBERG, J. H. and NOE, J. D.; Distributed amplification. *Proc. Inst. Radio Engrs.* 1948 **36** 956–69.
518. HORTON, W. H., JASBERG, J. H. and NOE, J. D.; Distributed amplifiers: practical considerations. *Proc. Inst. Radio Engrs* 1950 **38** 748–53.
519. STEINBERG, J. L.; On the theory of semi-distributed amplifiers. *Onde Elect.* (March) 1950 **30** 121–7.
520. WEBER, J.; Distributed amplification: additional considerations. *Proc. Inst. Radio Engrs* 1951 **39** 310.
521. COPSON, A. P.; A distributed power amplifier. *Electrical Engng (N.Y.)* 1950 **69** 893–8.
522. CORMACK, A.; Distributed amplification. *Electronic Engng* 1952 **24** 144–7.
523. KELLEY, G. G.; High speed synchroscope. *Rev. Sci. Instrum.* 1950 **21** 71–76.
524. KENNEDY, F. and RUDENBERG, H. G.; Wide-band chain amplifier. *Electrical Manufacturing* (Nov.) 1949.
525. MYERS, G. F.; Short pulse amplifier. *Electronics* (Jan.) 1952 **25** 128–31.

526. RUDENBERG, H. G. and KENNEDY, F.; 200 Mc/s. travelling-wave chain amplifier. *Electronics* (Dec.) 1949 **22** 106–9.
527. SCHARFMAN, H.; Distributed amplifier covers 10–360 Mc/s. *Electronics* (July) 1952 **25** 113–115.
528. TYMINSKI, W. V.; Wide-band chain amplifier for TV. *Radio-Electronic Engng* (April) 1950 **14** 14–6, 29.
529. YU, Y. P., KALLMAN, H. E. and CHRISTALDI, P. S.; Millimicrosecond oscillography. *Electronics* (July) 1951 **24** 106–11.
530. LEWIS, I. A. D.; Analysis of a helical transmission line with thermionic valve action. *A.E.R.E. Report* G/R 611, January 1950. Analysis of a transmission-line type of thermionic amplifier valve. *Instn Elect. Engrs Monograph* No. 57, December 1952; *Proc. Instn Elect. Engrs* 1953, Pt. IV.
531. FOWLER, V. J.; Transmission line tubes. *Proc. Nat. Electronics Conf.* 1951 **7** (*Electronics* January 1952 **25** 214–218).
532. FOWLER, V. J.; Transmission line tubes. University of Illinois, *Elec. Engng Res. Lab. Tech. Report* No. 15, September 1951.
533. PIERCE, J. R.; Travelling-wave tubes. Van Nostrand, 1950.
534. MATHEWS, W. E.; Travelling-wave amplification by means of coupled transmission lines. *Proc. Inst. Radio Engrs* 1951 **39** 1044–51.
535. ADLER, R.; Miniature travelling-wave tube. *Electronics* (Oct.) 1951 **24** 110–3.

Chapter 6—Cathode Ray Oscilloscopes

601. LEE, G. M.; A three beam oscillograph for recording at frequencies up to 10,000 Mc/s. *Proc. Inst. Radio Engrs* 1946 **34** (No. 3) 121 W.
602. ARDENNE, M. VON.; Der Elektronen Mikro-oscillograph. *Hochfrequenztechnik und Elektroakustik* 1939 **54** (No. 6) 181.
603. HOLLMAN, H. E.; The travelling-wave cathode ray tube. *Proc. Inst. Radio Engrs* 1951 **39** (No. 2) 194.
604. PIERCE, J. R.; Travelling wave oscilloscope. *Electronics* 1949 **22** (No. 11) 97.
605. OWAKI, K., TERAHATA, S., HEDA, T. and NAKAMURA, T.; The travelling-wave cathode ray tube. *Proc. Inst. Radio Engrs* 1950 **38** (No. 10) 1172.
606. SMITH, S. T., TALBOT, R. V. and SMITH, C. H.; Cathode ray tube for recording high speed transients. *Proc. Inst. Radio Engrs* 1952 **40** (No. 3) 297.
607. LANGMUIR, D. B.; Theoretical limitations of cathode ray tubes. *Proc. Inst. Radio Engrs* 1937 **25** (No. 8) 977.
608. MALOFF, I. G. and EPSTEIN, D. W.; Electron optics in television: see Bibliography.
609. MOSS, H.; The electron gun of the cathode ray tube. *J. Brit. Inst. Radio Engrs* 1945 **5** (No. 1) 10 and 1946 **6** (No. 3) 99.
610. LIEBMANN, G.; The image formation in cathode ray tubes and the relation of fluorescent spot size and final anode voltage. *Proc. Inst. Radio Engrs* 1945 **33** (No. 6) 381.
611. DUDDING, R. W.; Aluminium-backed screens for cathode ray tubes. *J. Brit. Inst. Radio Engrs* 1951 **11** (No. 10) 455.
612. PIERCE, J. R.; After Acceleration and Deflection. *Proc. Inst. Radio Engrs* 1941 **29** (No. 1) 28.
613. ALLARD, L. S.; An ideal post deflexion accelerator C.R.T. *Electronic Engng* 1950 **22** (No. 273) 461.
614. ROGOWSKI, W. FLEGLER, E. *und* BUSS, K.; Die Leistungsgrenze des Kathoden-oszillographen. *Arch. für Elektrotechnik* 1930 **24** 563.

615. HEROCK, R. J.; The photographic recording of cathode ray tube traces. See Bibliography.
616. HOWELLS, G. A.; Atomic Energy Research Establishment, Harwell, England. Private communication.
617. KELLEY, G. G; High Speed Synchroscope. *Rev. Sci. Instrum.* 1950 **21** (No. 1) 71.
618. MOODY, N. F. and McLUSKY, G. J. R.; Millimicrosecond Pulse Techniques (Part 2). *Electronic Engng* 1952 **24** (No. 292) 287.
619. BAUER, R. E. and NETHERCOT, W.; A new oscillograph for the recording of very fast electrical transients. *British Electrical and Allied Industries Research Association Technical Report.* U/T115 (1949).
620. SMITH, D. O.; A sweep system for the micro-oscillograph. Laboratory for Insulation Research, M.I.T. Cambridge, Mass. U.S.A. March 1950.
621. PRIME, H. A. and RAVENHILL, P.; The design of a high speed oscillograph. *J. Sci. Instrum.* 1950 **27** (July) 192.
622. WELLS, F. H.; Fast pulse circuit techniques for scintillation counters. *Nucleonics* 1952 **10** (No. 4) 28.
623. JANSSEN, J. M. L.; An experimental stroboscopic oscilloscope for frequencies up to about 50 Mc/s. *Philips Tech. Review* 1950 **12** (No. 2) 52 and 1950 **12** (No. 3) 73.
624. MCQUEEN, J. G.; The monitoring of high speed waveforms. *Electronic Engng* 1952 **24** (No. 296) 436.

Chapter 7—Applications To Nuclear Physics

701. COLTMAN, J. W.; The scintillation counter. *Proc. Inst. Radio Engrs* 1949 **37** (No. 6) 671.
702. POST, R. F. and SCHIFF, L. I.; Statistical limitations on the resolving time of a scintillation counter. *Phys. Rev.* 1950 **80** (No. 6) 1113.
703. COLLINS, G. B.; Decay times of scintillations. *Phys. Rev.* 1948 **74** (No. 10) 1543.
704. POST, R. F. and SHIREN, N. S.; Decay time of stilbene scintillations as a function of temperature. *Phys. Rev.* 1950 **78** (No. 1) 80.
705. LUNDBY, A.; Scintillation decay times. *Phys. Rev.* 1950 **80** (No. 3) 477.
706. BITTMAN, L. FURST, M. and KALLMANN, H.; Decay times, fluorescent efficiences and energy storage properties for various substances with gamma ray or alpha particle excitation. *Phys. Rev.* 1952 **87** (No. 1) 83.
707. ALLEN, J. S. and ENGELDER, T. C.; Scintillation counting with an E.M.I. 5311 photomultiplier tube. *Rev. Sci. Instrum.* 1951 **22** (No. 6) 401.
708. RAJCHMAN, J. A. and SNYDER, R. L.; An electrically focussed multiplier phototube. *Electronics* 1940 **13** (No. 12) 20.
709. RODDA, S.; *Photoelectric Multipliers.* *J. Sci. Instrum.* 1949 **26** (No. 3) 65.
710. GREENBLATT, M. H., GREEN, M. W., DAVISON, P. W. and MORTON, G. A.; Two new photomultipliers for scintillation counting. *Nucleonics* 1952 **10** (No. 8) 44.
711. SMITH, L. G.; Magnetic electron multipliers for detection of positive ions. *Rev. Sci. Instrum.* 1951 **22** (No. 3) 166.
712. WELLS, F. H.; Fast pulse circuit techniques for scintillation counters. *Nucleonics* 1952 **10** (No. 4) 28.
713. ČERENKOV, P. A.; Visible radiation produced by electrons moving in a medium with velocities exceeding that of light. *Phys. Rev.* (Aug. 15) 1937 **52** 378.
714. MARSHALL, J.; Particle counting by Čerenkov radiation. *Phys. Rev.* 1952 **86** (No. 5) 685.

715. GREINACHER, H.; Über einen hydraulischen Zähler für Elementarstrahlen. *Helvetica Physica Acta* 1934 **7** 360.
716. STUBER, R.; Über die Wirkungsweise des Funkenzählers. *Helvetica Physica Acta* 1939 **12** 109.
717. CHANG, W. Y. and ROSENBLUM, S.; A simple counting system for alpha-ray spectra and the energy distribution of Po alpha particles. *Phys. Rev.* 1945 **67** (Nos. 7 and 8) 222.
718. KEUFFEL, J. W.; Parallel Plate Counters. *Rev. Sci. Instrum.* 1949 **20** (No. 3) 202.
719. MADANSKY, L. and PIDD, R. W.; Some properties of the parallel plate spark counter II. *Rev. Sci. Instrum.* 1950 **21** (No. 5) 407.
720. ROBINSON, E.; Spark Counters for short time interval cosmic ray measurements. *Proc. Phys. Soc.* 1953 **66** (Pt. 1, No. 397A) 73.
721. MOODY, N. F.; Millimicrosecond pulse techniques (Pt. 1). *Electronic Engng* 1952 **24** (No. 291) 214.
722. HOWELLS, G. A.; Atomic Energy Research Establishment, Harwell England. Private communication.
723. WELLS, F. H.; A fast amplitude discriminator and scale of ten counting unit for nuclear work. *J. Sci. Instrum.* 1952 **29** (No. 4) 111.
724. FITCH, V.; A high resolution scale of four. *Rev. Sci. Instrum.* 1949 **20**. (No. 12) 942.
725. SESSLER, W. M. and MASKET, A. V.; High speed electronic scaler. *Rev. Sci. Instrum.* 1950 **21** (No. 5) 494.
726. BELL, R. E., GRAHAM, R. L. and PETCH, H. E.; Design and use of coincidence circuit of short resolving time. *Canad. J. Phys.* 1952 **30** (No. 1) 35.
727. WELLS, F. H.; Pulse circuits for the millimicrosecond range. *Brit. Inst. Radio Engrs.* 1951 **11** (No. 11) 491.
728. BEGHIAN, L. E., ALLEN, R. A., CALVERT, J. M. and HALBAN, H.; A fast neutron spectrometer. *Phys. Rev.* 1952 **86** (No. 6) 1044.
729. MCLUSKY, G. R. J. and MOODY, N. F.; Millimicrosecond pulse techniques (Pt. 4). *Electronic Engng* 1952 **24** (No. 293) 330.
730. BAY, Z., CLELAND, M. R. and MCLERNON, F.; Fast coincidences with Čerenkov counters. *Phys. Rev.* 1952 **87** (No. 5) 901.
731. BAY, Z.; A new type of high-speed coincidence circuit. *Rev. Sci. Instrum.* 1951 **22** (No. 6) 397.
732. BALDINGER, E., HUBER, P. and MEYER, K. P.; High-speed coincidence circuit used for multipliers. *Rev. Sci. Instrum.* 1948 **19** (No. 7) 473.
733. ELLMORE, W. C.; Coincidence circuit for a scintillation detector of radiation. *Rev. Sci. Instrum.* 1950 **21** (No. 7) 649.
734. DICKE, R. H.; A high-speed coincidence circuit. *Rev. Sci. Instrum.* 1947 **18** (No. 12) 907.
735. FISCHER, J. and MARSHALL, J.; The 6BN6 gated beam tube as a fast coincidence circuit. *Rev. Sci. Instrum.* 1952 **23** (No. 8) 417.
736. HOFSTADTER, R. and MCINTYRE, J. A.; Note on the detection of coincidences and short time intervals. *Rev. Sci. Instrum.* 1950 **21** (No. 1) 52.
737. GARWIN, R. L.; A useful fast coincidence circuit. *Rev. Sci. Instrum.* 1950 **21** (No. 6) 569.
738. BAY, Z. and PAPP, G.; Coincidence device of 10^{-8} to 10^{-9} second resolving power. *Rev. Sci. Instrum.* 1948 **19** (No. 9) 565.
739. LUNDBY, A.; Delayed coincidence circuit for scintillation counters. *Rev. Sci. Instrum.* 1951 **22** (No. 5) 324.

740. CLELAND, M. R. and JASTRAM, P. S.; The velocity of gamma rays in air. *Phys. Rev.* 1951 **84** (No. 2) 271.
741. LUCKEY, D. and WEIL, J. W.; Velocity of 170 MeV gamma rays. *Phys. Rev.* 1952 **85** (No. 6) 1060.
742. NEWTON, T. D.; Decay constants from coincidence experiments. *Phys. Rev.* 1950 **78** (No. 4) 490.
743. BAY, Z.; Calculation of decay times from coincidence experiments. *Phys. Rev.* 1950 **77** (No. 3) 419.
744. BAY, Z., MEIJER, R. R. and PAPP, G.; On measuring very short half-lives. *Phys. Rev.* 1951 **82** (No. 5) 754.
745. KEUFFEL, J. W.; A simplified chronotron-type timing circuit. *Rev. Sci. Instrum.* 1949 **20** (No. 3) 197.
746. NEDDERMEYER, S. H., ALTHAUS, E. J., ALLISON, W. and SCHATZ, E. R.; The measurement of ultra-short time intervals. *Rev. Sci. Instrum.* 1947 **18** (No. 7) 488.
747. MOODY, N. F.; Millimicrosecond pulse techniques (Pt. 3). *Electronic Engng* 1952 **24** (No. 292) 289.
748. POST, R. F. and SHIREN, N. S.; Performance of pulsed photo-multiplier. *Phys. Rev.* 1950 **78** (No. 1) 81.

Chapter 8—Miscellaneous Applications

801. DELANGE, O. E.; Propagation studies at microwave frequencies by means of very short pulses. *Bell System Tech. Journal* 1952 **31** (No. 1) 91.
802. ENGLISH, W. N.; Photon pulses from point to plane corona. *Phys. Rev.* 1950 **77** (No. 6) 850.
803. ZAREM, A. M., MARSHALL, F. R. and POOLE, F. L.; An electro-optical shutter for photography. *Electrical Engng* 1949 **68** (No. 4) p. 282.
804. DUNNINGTON, F. G.; The electro-optical shutter—its theory and technique. *Phys. Rev.* 1931 **38** (No. 8) 1506.
805. ANDERSON, W. C.; Measurement of the velocity of light. *Rev. Sci. Instrum.* 1937 **8** (No. 7) 239.
806. FROOME, D. K.; Fundamental studies of the cathode spot formation in the arc discharge. *J. Sci. Instrum.* 1948 **25** (No. 11) 371.
807. QUINN, H. F., MCKAY, W. B. and BOURQUE, O. J.; A Kerr cell camera and flash illumination unit for ballistic photography. *J. Appl. Phys.* 1950 **21** (No. 10) 995.
808. BEAMS, J. W. and MORTON, H. S.; Transmission line Kerr cell. *J. Appl. Phys.* 1951 **22** (No. 4) 523.
809. HOLST, G.; British Patent No. 326,200.
810. JENKINS, J. A. and CHIPPENDALE, R. A.; The application of image converters to high speed photography. *J. Brit. Inst. Radio Engrs* 1951 **11** (No. 11) 505.

INDEX

Aluminizing, cathode ray tube screen 192
Amplifiers, figure of merit 135
— interstage coupling 138
— limitations of valves at high frequencies 133
— secondary emission valves 141
— transmission line tubes 167
— travelling-wave tubes 168
— *also see* Distributed amplifiers
Amplitude bandwidth 6
— discriminator circuits 206, 232
— discrimination in coincidence circuits 253
— distortion 6
Attenuators 121
— coaxial 124
— lossy transmission line 126
— simple 106, 122

Bandwidth of cathode ray tube deflecting plates 177
Beam acceleration after deflection in cathode ray tubes 193
Blocking oscillators 120, 128, 130, 242
Boot-strap time-base circuit 198
Brightness of cathode ray tube display 191

Cathode ray tube design 170
    aluminizing of screen 192
    beam acceleration after deflection 193
    brightness of display 191
    deflecting plate connections 174, 177
    electron gun voltage, effect on spot size 189
    micro-oscillograph technique 182
    multiple plate deflecting systems 182
    performance of some tubes 196
    spot size, effect on dimensions of display 191
    — — relation to deflection sensitivity 187
    time-base waveform distortion 180
    transit-time limitations 171, 177

Cathode ray tube design (*contd.*)
    travelling-wave deflecting plates 186
    unsealed cathode ray tubes 194
Čerenkov radiation 229
Chronotron timing unit 269
Circuits with distributed parameters 15
Coincidence circuits 244
— — amplitude discrimination 253
— — analysis of delayed coincidence curves 267
— — cathode ray tube mixing circuit technique 263
— — crystal diode mixing circuits 259
— — delayed coincidence circuits 264
— — double grid valve mixer 263
— — double pulse mixer technique 261
— — multiple delayed coincidence circuits 269
— — pulse limited by diode mixer 246
— — resolving-time, limitations 256
— — — — stability of 255
Constant current generator 2
— voltage generator 2

Deflection sensitivity of cathode ray tubes 187
Delay circuits for transient recording oscilloscope 197
Delay lines (lumped) 52
— — constant-*k* filters 53
— — derived filters 55, 151
Delayed coincidence circuits 264
Differentiating circuit 12
Distributed amplifiers 143
— — alternative couplings 146
— — complete circuits 160, 161, 165
— — derived filter networks 151
— — gain, stability of 157
— — high frequency limitations 155
— — low frequency limitations 152
— — noise 158
— — optimum grouping of valves 146
— — output stage 149
— — signal amplifier (details) 161

INDEX

Distributed amplifiers, 400 Mc/s amplifier (details) 165
Duhamel integral 8

Electrical discharge measurements 277
Electro-magnetic theory 16, 17, 25, 283
Electron multiplier, with crossed field focussing 225
— — also see Photomultipliers
Electro-optical shutters for high-speed photography 278

Filters 52
— constant-k 53
— coupling in amplifiers 138
— derived 55, 151
Fourier analysis 4, 9
— spectra 4, 5
Fourier-Mellin inversion integral 11

Half-lifetime measurements 266
Helical lines 42
— — cable data 52
— — capacitance per unit length 47
— — inductance per unit length 42
— — phase distortion 48
— — self-capacitance 50
— — variation of inductance with frequency 48

Image converter 280
Integrating circuit 12

Kerr cell 278

Laplace transform analysis 5, 9
— — — correspondence with sinusoidal analysis 11
— — — standard forms 10
Lee's three beam micro-oscillograph 182

Micro-oscillograph techniques 182

Oscilloscopes 170
— for recurrent waveform display 207
— for transient waveform recording 196
— using pulse sampling techniques 208
— also see Cathode ray tube design

Phase bandwidth 6
— distortion 6
— — in helical lines 48
— — in lumped delay lines 53, 55, 56

Phosphor decay time-constants 221, 223
Photographic equipment for transient recording oscilloscope 207
— technique of cathode ray tube recording 195
Photomultipliers for scintillation counters 224
Pulse distortion 6
— — in tapered lines 73
— — also see Phase distortion
— generators 99
— — discharge line type 100
— — further possibilities 120
— — photomultiplier testing 229
— — relays 102
— — secondary emission valves 115
— — tapered discharge line 112
— — thyratron 107
— inverter 91, 111
— rise-time 6
Pulse sampling techniques for recurrent waveform oscilloscope 208
  bandwidth 212
  brightness of display 217
  display circuit 215
  mixing circuits 210
  signal deflection sensitivity 217
Pulsed valve time-base circuits 202
— X-ray tubes 230

Radar propagation measurements 277
Radio receiver measurements 276
Recording oscilloscopes 196, 274
Relays 102, 107
Resistors, effects at high frequencies 121

Scaling circuits 239
Scintillation counters 219
— — current pulse shape 220
— — output circuits 225
— — performance 223
— — phosphor decay time 221
— — transit-time effects 224
Sinusoidally varying currents 2, 9
Skin effect 22, 39
Spark counters 231
Spot size, of cathode ray tube 187
Step function 7, 175, 213
Superposition, principle of 1, 5, 8

Thevenin's theorem 1, 4, 9
Thyratron pulse generators 107
— — time-base circuits 203
Time-base circuits 198

Time-base distortion in cathode ray tubes 180
Time calibration circuits for oscilloscopes 206
— expansion circuit 271
— interval measurements with coincidence circuits 264
— — — by integration techniques 271
— sorters using delayed coincidence circuits 269
Transformers 59
— coupled lines 96
— exponential lines 79, 113
— Gaussian line 64, 76, 90
— linearly tapered line 90
— lumped 61, 130
— other laws of impedance variation 91
— tapered lines 63
Transient recording oscilloscope circuit design 193
Transit-time effects in cathode ray tubes 171, 177
— — — — photomultipliers 244
Transmission lines as circuit elements 32
    input impedance 36
    pulse shaping 34
    voltage at input end 33, 37

Transmission lines, arbitrary termination 22, 26
— — coaxial cable data 41
— — current reflection coefficient 24
— — — transmission coefficient 24
— — discontinuities 22
— — — capacitance equivalents 27
— — general properties of uniform lines 15, 18
— — higher modes 25
— — losses, *also see* Attenuators 38
— — matching 31, 60, 106
— — pulse inverter 91, 111
— — standing-wave ratio 25
— — supports 29
— — tapered, *see* Transformers
— — voltage reflection coefficient 23, 24
— — — transmission coefficient 24
— — *also see* Helical lines
Transmission-line tubes 167
Travelling-wave deflecting systems, multiple plates 182
— — — — distributed constants 186
Trigger amplifiers for time-base circuits 205

Valve voltmeter, reflex peak 127
Variable circuit 14